New Drugs Targeting Antibiotic-Resistant Bacteria: Recent Advances

Edited by

Mariano Martínez-Vázquez
Department of Natural Products
Chemistry Institute, National Autonomous University of
México
México City, México

New Drugs Targeting Antibiotic-Resistant Bacteria: Recent Advances

Editor: Mariano Martínez-Vázquez

ISBN (Online): 979-8-89881-240-9

ISBN (Print): 979-8-89881-241-6

ISBN (Paperback): 979-8-89881-242-3

need for a court order if at any point you breach any terms of this License Agreement. In no event will any delay or failure by Bentham Science Publishers in enforcing your compliance with this License Agreement constitute a waiver of any of its rights.

3. You acknowledge that you have read this License Agreement, and agree to be bound by its terms and conditions. To the extent that any other terms and conditions presented on any website of Bentham Science Publishers conflict with, or are inconsistent with, the terms and conditions set out in this License Agreement, you acknowledge that the terms and conditions set out in this License Agreement shall prevail.

Bentham Science Publishers Pte. Ltd.
No. 9 Raffles Place
Office No. 26-01
Singapore 048619
Singapore
Email: subscriptions@benthamscience.net

BENTHAM SCIENCE

CONTENTS

PREFACE

Bacterial strains' resistance to bactericides is a natural process that arises from the selection pressure of antibiotics against these pathogens. The continuous assault of drugs on bacteria prompts these organisms to develop various defense mechanisms, such as β-lactamase biosynthesis, which targets the hydrolysis of the β-lactam ring in different penicillin and carbapenems. Other resistance mechanisms exist, such as changes in cell membrane proteins or the increased activity of efflux pumps. These changes are derived from gene information from surviving bacteria to pharmacological treatments. The dissemination of this information is done very efficiently both by vertical and transverse routes. It is known that penicillin resistance emerged only a few years after the implementation of penicillin as a worldwide antibacterial drug. This was the beginning of a recurring phenomenon: a new drug was synthesized to treat a complex infection, and a time later, strains were resistant to this new medication. The misuse of antibiotics in livestock production and among humans, where antibiotic medication was unnecessary, has accelerated the emergence of Multidrug-Resistant Strains (MDR). Upon reaching concerning levels of bacterial resistance worldwide in 2017, the World Health Organization launched a list called ESKAPE, which identifies the main resistant drug bacteria that constitute a threat and require new drugs with new routes of action. Recently, the Centers for Disease Control and Prevention (CDC-USA), considering the rise of infections and deaths induced by microorganisms, classified the infections as urgent, serious, and concerning threats. It is worth noting that the CDC list differs from that proposed by the World Health Organization in 2017. As Chapter One indicates, antibiotic resistance is one of humanity's most immense health challenges. Virtually any bacterium can develop resistance, either by its evolution, by antibiotic pressure, or by genetic exchanges. Additionally, in this chapter, the author emphasizes the need for a one-health approach, addressing MDRO comprehensively in humans, animals, soil, water, and manure. The focus is on infection prevention and control, as well as optimizing antibiotic use to break the chain of resistance acquisition and transmission. In chapter two, the authors cover various aspects of phage therapy, including isolating and characterizing bacteriophages, the development of suitable formulations, and their administration for treating human infections. It also examines successful cases of phage therapy in treating life-threatening infections that cannot be cured with current antibiotics. Moreover, the chapter highlights the advantages and limitations of phage therapy compared to traditional antimicrobials. Chapter three focuses on the ongoing efforts and strategies to find new antibacterial compounds from natural sources. It discusses traditional methods and innovative techniques for dealing with non-culturable microorganisms. The positive outcomes of these approaches offer hope for a potential resurgence in discovering these valuable molecules, marking a second golden age for the field. As indicated in Chapter Four, clinicians have limited options when choosing antibiotics. The only available option for them is carbapenems. To develop new pharmacological strategies, clinicians must identify enzymes that break down these antimicrobials. Identifying these enzymes is crucial to reduce the selection pressure and ensure the correct use of antibiotics. The length of time a molecule stays active in the body is determined by its persistence. Clinical microbiology laboratories play a crucial role in guiding the administration of drugs. This chapter discusses the importance of carbapenemases-mediated resistance, its classification, impact, and detection strategies. Chapter Five identifies the most effective bactericidal-drug combinations against drug-resistant bacteria. Combining the carbapenem antibiotic and a β-lactamase inhibitor is considered one of the best combinations for clinical treatments. The chapter also discusses the use of Quorum-sensing inhibitors to inhibit virulence. Research on virulence inhibitors based on halogen furanone-type compounds has shown efficient virulence inhibition *in vitro*. However, there are currently no QS inhibitors in clinical evaluation.

Additionally, the synthesis of nanoparticles to counteract drug-resistant bacteria is explored. Nanoparticles synthesized with biological activity have shown significant results, especially those made from metals like silver and those synthesized using polymeric materials with biodegradable substances. Further studies are needed to determine their effectiveness and toxicity.

The editor would like to thank all the authors for their dedication and time in creating this book. He also thanks Ms. Graciela Flores-Rosete for compiling and organizing all the chapters.

Mariano Martínez-Vázquez
Department of Natural Products
Chemistry Institute, National Autonomous University of México
México City, México

List of Contributors

Daniela de la Rosa Zamboni	Subdirectorate of Comprehensive Patient Care, Children's Hospital of México, México City, México
Daniela Luis-Yong	Center for Research in Applied Mycology, Veracruzana University, Veracruz, México
Daniel Huelgas-Mendez	Microbiology and Parasitology Department, Faculty of Medicine, National Autonomous University of México, México City, México
Israel Castillo-Juárez	Institute of Basic Sciences and Engineering, Secihti-Autonomous University of the State of Hidalgo, Hidalgo, México
Jorge Santiago Jimenez-Zuñiga	Microbiology and Parasitology Department, Faculty of Medicine, National Autonomous University of México, México City, México
José Luis Díaz-Nuñez	Postgraduate in Botany, Comecyt–College of Postgraduates, México
Juan Carlos García-Cruz	Microbiology and Parasitology Department, Faculty of Medicine, National Autonomous University of México, México City, México
Luis Esau López Jacome	Clinical Microbiology Laboratory, Infectious Diseases Division, Infectious Diseases Division, National Institute of Rehabilitation Luis Guillermo Ibarra Ibarra, México City, México Biology Department, National Autonomous University of México, México City, México
Mariel Hernández-Garnica	Microbiology and Parasitology Department, Faculty of Medicine, National Autonomous University of México, México City, México
Mariano Martínez Vázquez	Microbiology and Parasitology Department, National Autonomous University of México, MéxicoCity, México
Rafael Franco Cendejas	Subdirection of Biomedical Research, National Autonomous University of México, México City, México
Rodolfo García-Contreras	Microbiology and Parasitology Department, Faculty of Medicine, National Autonomous University of México, México City, México
Xareni Rebollar Juarez	Microbiology and Parasitology Department, Faculty of Medicine, National Autonomous University of México, México City, México

CHAPTER 1

Multidrug Resistance, An Unmeasurable Epidemic (MDRO)

Daniela de la Rosa Zamboni[1,*]

[1] *Comprehensive Patient Care Department, Children's Hospital of México, México City, México*

Abstract: Despite the undeniable benefits of antibiotics, the emergence of resistance presents a formidable challenge. This section examines the evolution of Multidrug-Resistant Organisms (MDROs), highlighting the complexities of systematic reporting and advocating for comprehensive surveillance to understand the true extent of the epidemic.

Initial successes of antibiotics, exemplified by penicillin, were short-lived as resistance swiftly emerged, marking the onset of the antibiotic resistance era. The current scenario reveals elevated resistance rates, particularly in low- and middle-income countries.

Medical concerns and reports highlight the ongoing apprehensions within the medical community regarding multidrug resistance, which dates back to the 1960s. Recent reports emphasize a global crisis of antibiotic scarcity and the rapid development of resistance. Livestock, pets, and other animals, as well as water and vegetables, are also contributing to the MDRO epidemic. The involvement of environmental animals and vegetables emphasizes the need for active epidemiological surveillance across all these sectors to prevent the transmission of MDRO, reinforcing the importance of environmental sanitation.

In conclusion, the origin and extent of the MDRO epidemic remain challenging to determine despite recent global surveillance efforts. The impact on health indicates a high association with mortality. There is a need for a One Health approach, addressing MDRO comprehensively in humans, animals, soil, water, and manure. The focus is on infection prevention and control, as well as optimizing antibiotic use to break the resistance acquisition and transmission chain.

Keywords: Antibiotic resistance, Epidemiological surveillance, Multidrug-resistant organisms.

---

[*] **Corresponding author Daniela de la Rosa Zamboni:** Comprehensive Patient Care Department, Children's Hospital of México, México City, México; E-mail: rzdaniela@hotmail.com

Mariano Martínez-Vázquez (Ed.)

INTRODUCTION

Diseases with high prevalence, incidence, and mortality, especially if they are contagious, typically require epidemiological surveillance. They have established and generally standardized reporting systems. However, there is an epidemic that started several decades ago. Despite its high prevalence and lethality, antibiotic resistance does not have mandatory reporting in every country, at least not in a generalized and extended manner. Antibiotic resistance has been documented for over a century, almost simultaneously with the discovery of some drugs that have prevented the most deaths, such as antibiotics.

Antibiotics have prevented millions of deaths and are regarded as one of the most significant discoveries in healthcare. Furthermore, critical medical advances, such as organ transplants, implants, cancer treatments, and complex surgeries, would not have been possible without antibiotics. The benefits and medical achievements made possible by antibiotics are undeniable. Additionally, the use of antibiotics is widespread in veterinary medicine, as well as in the cattle, poultry, and swine industries, and even agribusiness.

Antibiotic resistance is one of humanity's most significant health challenges. Virtually any bacterium can develop resistance through evolution, antibiotic pressure, or genetic exchange [1 - 4]. Infections caused by resistant bacteria are difficult or impossible to treat, often require complex microbiological or molecular technology for diagnosis, and are generally more lethal than their antibiotic-sensitive counterparts [5, 6]. Furthermore, antibiotic resistance affects not only humans but also animals and plants. Therefore, antibiotic resistance has been complex to study, analyze, prevent, and control.

Antibiotic resistance has gradually and continuously evolved at an alarming rate. There are multiple reports of drug resistance worldwide, and despite being local or partial reports, it is evident that the resistance epidemic has progressed to multidrug resistance, with its incidence on the rise. Unfortunately, recent systematic review studies or studies that estimate the global burden of disease due to multidrug resistance indicate that antibiotic resistance is ranked third in deaths worldwide [7].

The source of MDRO is not just the use of antibiotics in humans and poor infection control in hospitals. The use of antibiotics in chickens, cattle, and pigs, as well as the use of manure as a fertilizer for plants, plays an important role, but their reporting is poorly standardized [8].

However, with the growing epidemic, alert MDRO reporting is not yet subject to standardized epidemiological surveillance as a cause of death. Although reported by researchers with extensive international mathematical models, the reporting of data about the implications of morbidity and mortality is poorly standardized. The low perception of damage, the lack of diagnostic methods, and the little interaction between the participants in its genesis, i.e., doctors, antibiotic prescribers, veterinarians, farmers, chicken breeders, pigs, and even farmers, among other factors [8, 9], are some of the factors that could contribute to the lack of measurement of the absolute magnitude of the epidemic.

This chapter provides the reader with a general overview of the emergence and widespread evolution of MDROs, including notions about the importance and difficulty of systematic and mandatory reporting and the extension of the epidemic in humans and beyond.

THE BEGINNING AND EVOLUTION OF THE PANDEMIC

Antibiotics have saved millions of lives since their systematic use; they are undoubtedly a remarkable achievement in the fight against bacteria [1]. However, the evolution of bacteria towards resistance has been so rapid that it has exceeded the discovery of new antibiotics. It is calculated by mathematical modeling studies that currently, deaths associated with antibiotic resistance are among the most frequent, only surpassed by stroke and ischemic heart disease [5, 6].

It is indisputable that antibiotic resistance is a natural aspect of microorganism evolution and that resistance to other organisms occurred before the widespread use of antibiotics. Some microorganisms have developed mechanisms to resist antibiotics; therefore, they produce antibiotics themselves [1, 2]. The first antibiotics used were substances produced by resistant microorganisms to kill other microorganisms [1, 2].

Since Hippocratic medicine, antibiotics have been used empirically. The empirical use of baker's yeast to treat wounds has been documented for over two thousand years in Egypt, Greece, and Serbia. About a thousand years ago, a recipe for plant-based eye drops with activity against Staphylococcus was found. This eye drop was recently recreated, and its effectiveness was demonstrated [1, 2, 11].

In 1871, Joseph Lister discovered that the fungus *Penicillium brevicompactum* inhibited bacterial growth. The extract was used to treat an infected wound on a nurse in 1893, hence the hypothesis that bacteria caused the infection. Bartolomeo Gosio isolated mycophenolic acid in 1893, which he used to treat Anthrax. Its use was forgotten until 1912, when it was revived by C.L. Alsberg and O.M. Black, who discovered it inhibited mitosis. Mycophenolic acid was forgotten until its immunosuppressive properties were later found [12 - 14]. It is currently known that there may be resistance to Mycophenolic Acid by a type of Candida, although its use to treat infections caused by it is not widespread [10, 12 - 14].

In 1910, Paul Ehrlich discovered the arsenic prodrug (arsphenamine), known as Salvarsan, which was used to treat syphilis. In 1913, it evolved into Neosalvarsan, which was less dangerous and more effective for treating syphilis. This marked the beginning of synthetic antibiotics [2]. Later, a broad-spectrum sulfonamide called Prontosil emerged in 1930, initially used mainly for treating soldiers during the First World War and, in 1936, for puerperal fever [3]. However, shortly after its use, mutations in the enzyme dihydropteroate synthetase—the target of the antibiotic—rendered it ineffective, leading to its replacement by penicillin, which was discovered in 1928 by Edward Fleming [2]. The structure of penicillin was elucidated in 1939 by Howard Walter Florey and Ernst Boris Chain, allowing for its purification and large-scale production. In 1945, the use of penicillin began [4]. Although penicillin was considered miraculous at the time, resistance soon emerged; in fact, penicillin resistance was identified after its discovery but before its approval for clinical use. Almost simultaneously with the advent of penicillin, the "golden age" of antibiotics commenced [5]. From the 1940s to the 1960s, several classes of antibiotics were discovered. These include antibiotics derived from actinomycetes (Aminoglycosides, Tetracyclines, Amphenicos, Macrolides, Glycopeptides, Tuveractinomycins, Ansamycins, Lincosamides, Streptogramins, Cycloserine, and Phosphonates), bacterial natural products (Polypeptides, Bacitracin, Polymyxins), natural products (Fusidic Acid, Cephalosporins, Enniatins), and synthetic antibiotics (Sulfones, Salycylates, Nitrofurans, Pyridinamides, quinolones, Azoles, Phenazines, Diaminopyrimidines, Ethambutol, and Thionamides). Unfortunately, bacterial resistance appeared shortly after the discovery of these antibiotics, in some cases even in the same year. Methicillin-resistant Staphylococcus aureus and plasmid-borne resistance to sulfonamides were detected in the early sixties [5].

After this "golden era," the discovery of antibiotics became scarce. From the '70s to the 2000s, fewer antibiotics were discovered, including phosphonates, carbapenems, mupirocin, and monobactams in the '80s, and lipopeptides, pleuromutilin, and oxazolidinones in 2000 [5].

In the last twenty years, fewer than a dozen new antibiotics and no new types have been discovered. However, antibiotic resistance continues to advance. The World Bank estimates that 39 to 70% of *E. coli* are currently resistant to co-trimoxazole, and a significant percentage of MRSA are resistant to Oxacillin [5].

MEDICAL CONCERNS AND REPORTS

In the 1960s, the first reports possibly indicating concern by the medical profession about the potential consequences of multidrug resistance were documented. Gynecologists and general practitioners, among others, published papers, issued communiqués, and delivered conferences on the subject [3, 15]. In the early 21st century, with the alarming title "bad bugs, no drugs, no ESKAPE," marking a series of reports, some international organizations, such as the Infectious Diseases Society of America and the CDC, reported the shortage of antibiotics and the rapid development of resistance. They also recognized it as a global crisis with an undefined magnitude [16, 17].

Finally, in 2015, the first global epidemiological surveillance system was initiated. This year, the WHO initiated a system for epidemiological surveillance of bacterial resistance, known as GLASS, which stands for Global Antimicrobial Resistance and Use Surveillance System. Subsequently, the consumption of antimicrobials and antifungals was monitored, and a system was implemented that considered the health of people, animals, and the environment. It currently has information from 127 countries and is in the initial implementation phase [18].

According to GLASS, over the last five years, antibiotic resistance has remained stable or increased slightly, but the resistance of a few bacteria has decreased. However, rates of antimicrobial resistance are high, particularly in low- and middle-income countries [18]. Antibiotic resistance poses alarming proportions, especially in low- and middle-income nations. The proportion of Escherichia coli infections resistant to third-generation cephalosporins has a median of 58.3% (39.8 to 70%) in low- and middle-income countries and 17.5% (11.3 to 25.21%) in high-income countries [3]. For methicillin-resistant Staphylococcus aureus, the numbers are 33.3% (19.5 to 55.6%) and 15% (6.8 to 36.4%), respectively. Both resistance profiles serve as global indicators of sustainable development linked to Sustainable Development Target 3.d ("strengthen the capacity of all countries, in particular developing countries, for early warning, risk reduction, and disaster management") [18].

In 2022, one of the first global epidemiological approaches to analyze the impact of MDRO on mortality was conducted. The article explores all the information available so far, covering nearly every country worldwide. It shows that MDRO is associated with the third leading cause of death globally, accounting for approximately 4.5 million deaths [5]. As a direct cause of death, it ranks among the top 15 causes of death. Once again, low- and very low-income countries experience a greater impact than high-income countries [5]. The figures reported in this article may be underestimated, as independent reports were considered from countries that, in most cases, were not standardized.

LIVESTOCK, PETS, AND OTHER ANIMALS, AND THE EPIDEMIC OF MDRO

Antibiotics have been used widely in chickens and cattle for several decades. Antibiotic resistance has since been found in meat. Manure, commonly used for fertilization, also contains resistant bacteria. As a result of this resistance, 97% of the resistance to ciprofloxacin in *E. coli* was found in humans [5, 19].

The effects of the use of antibiotics in animals are found not only in manure but also in food-borne infections, most of which are caused by consuming contaminated meat products [5, 19].

The main bacterial agents that cause foodborne diseases—Salmonella, Listeria, Staphylococcus, E. coli, and Campylobacter—have been reported to exhibit bacterial resistance in some countries, such as Bangladesh, reaching up to 97% [8]. However, in the United Kingdom, fluoroquinolone resistance is predicted to be 75% by 2040 [20, 21].

Regarding vegetables and fruits, resistant bacteria have also been found, which genotypically coincide with the bacteria in irrigation water or manure with which they are fertilized. Outbreaks associated with the consumption of vegetables have identified irrigation water, manure, or handling during storage, packaging, transportation, or distribution as the source [8, 22].

Although not frequent, contamination has been found from the same sources as contamination by reptiles or amphibians, such as Salmonella, and outbreaks involving pets infected with Campylobacter [22, 23]. Lastly, antibiotics are widely used in aquaculture and the ornamental fish industry. However, the role of this use or the fish in transmitting MDRO is unknown [22].

"WATERS."

Drainage water may also contribute to the epidemic, particularly from hospitals or farms. Sewage from these facilities has been shown to contain a higher concentration of multidrug-resistant bacteria than water from other sources. Similarly, reports indicate MDRO disease outbreaks originating from such sites. Environmental bacteria exhibit a greater tendency for genetic transmission than their non-environmental counterparts. The role of sewage in disease transmission still necessitates further investigation. Nevertheless, active epidemiological surveillance of MDROs in wastewater could be crucial for quantifying and preventing the transmission of MDROs. It is clear that environmental sanitation, particularly water treatment and safe disposal of excreta, is vital for avoiding MDRO [22].

Wastewater can reach rivers and streams, where it may cause contamination. In Mexico, in the Lerma River, one of the country's main rivers, bacteria of the ESKAPE group were found to be resistant in up to 15% of the samples [25].

RECREATIONAL WATERS

Few reports exist about the transmission of MDRO through recreational water use. However, recreational water use is linked to community outbreaks of resistant enterobacteria, with a more than doubled risk in individuals who use swimming pools. The risk appeared higher when recreational water was located near a farm [22 - 26]. The recreational use of seawater is also related to an increased risk of carrying resistant enterobacteria. The type of activity seems to influence this risk. For example, when comparing surfers to non-surfers, the former had a colonization percentage of 6.5% compared to 1.5% for non-surfers. Surfers ingest 150 ml or more during their activity, while non-surfers consume only around 30 ml [26].

THE NEED FOR A SYSTEMATIC AND MANDATORY REPORT OF MDRO AND THE BURDEN OF DISEASE

Diseases of high epidemiological risk are those that are subject to epidemiological surveillance. In the world, they have established and generally standardized reporting systems. Although, in general, it is considered that there is an underreporting, the diseases subject to epidemiological surveillance, among which are HIV - AIDS, malaria, tetanus, and neural tube defects, have tools for diagnostic scrutiny and standardized reporting or are in the process of being standardized. Countries also have healthcare quality indicators that rarely include diseases subject to epidemiological surveillance.

However, the MDRO epidemic began less than a decade ago with global reporting, which tends to be systematized, limited to reports of bacterial resistance and the use of antibiotics in humans, primarily by hospitals.

Despite its impact, antibiotic use in the agro-industry is under little surveillance. The resistance generated by antibiotics in agribusiness is even more poorly determined [22, 23].

MDRO infections, despite their relevance and frequency, are not yet included in the ICD-10 catalog, so the actual impact on mortality, morbidity, and even costs is difficult to calculate.

Systematic mandatory reporting can lead to national or international policies for effective control measures. On the contrary, if it continues as a diagnosis that is the subject of research, it will be challenging to know and will impact the actual impact of the disease.

Hospitals, health providers, veterinarians, the poultry industry, livestock and agriculture, aquaculture, and especially the government, must establish the necessary mechanisms at the national or international level to address multidrug resistance and notify authorities of any infectious disease outbreaks.

CONCLUSION

The epidemiological origin and extent of the multidrug-resistant epidemic are difficult to determine. Until recently, several centuries after the first bacterial resistance was found, epidemiological surveillance has been extended worldwide with the GLASS report. The impact on health is not yet clear, although it seems to indicate that at least the third place in deaths is associated with MDRO, and MDRO as a direct cause of death occupies the first place in mortality. Unlike other notifiable diseases, MDRO still has few indicators, and its study must be extended to human beings, cattle, chickens, pets, soil, manure, and water.

To address the epidemiologic impact of MDRO, it is essential to extend efforts toward a health approach with a strong emphasis on infection prevention and control, as well as antibiotic optimization, to disrupt the chain of resistance acquisition and MDRO transmission.

REFERENCES

[1]   Ribeiro da Cunha B, Fonseca LP, Calado CRC. Fonseca LP, Calado CRC. Antibiotic discovery: where have we come from, where do we go? Antibiotics (Basel) 2019; 8(2): 45.
[http://dx.doi.org/10.3390/antibiotics8020045] [PMID: 31022923]

[2]   Barber M. The use and abuse of antibiotics. BJOG 1960; 67(5): 727-32.
[http://dx.doi.org/10.1111/j.1471-0528.1960.tb10425.x]

[3]   Gaynes R. The discovery of penicillin—new insights after more than 75 years of clinical use. Emerg Infect Dis 2017; 23(5): 849-53.
[http://dx.doi.org/10.3201/eid2305.161556]

[4]   Hutchings MI, Truman AW, Wilkinson B. Antibiotics: past, present and future. Curr Opin Microbiol 2019; 51: 72-80.
[http://dx.doi.org/10.1016/j.mib.2019.10.008] [PMID: 31733401]

[5]   Rafiq K, Islam MR, Siddiky NA, *et al.* Antimicrobial resistance profile of common foodborne pathogens recovered from livestock and poultry in Bangladesh. Antibiotics (Basel) 2022; 11(11): 1551.

[http://dx.doi.org/10.3390/antibiotics11111551] [PMID: 36358208]

[6]   Ranjbar R, Alam M. Antimicrobial Resistance Collaborators (2022). Global burden of bacterial antimicrobial resistance in 2019: a systematic analysis. Evid Based Nurs 2024; 27(1): 16-6.
[http://dx.doi.org/10.1136/ebnurs-2022-103540] [PMID: 37500506]

[7]   Initiatives for Addressing Antimicrobial Resistance in the Environment: Current Situation and Challenges 2018. https://wellcome.ac.uk/sites/default/files/antimicrobial-resistance-environme-t-report.pdf

[8]   Mahindroo J, Narayan C, Modgil , *et al.* Antimicrobial resistance in food-borne pathogens at humana nimal interface: Results from a large surveillance study in India. One Healt 2024; 18: 100677.
[http://dx.doi.org/10.1016/j.onehlt.2024.100677] [PMID: 39010970]

[9]   Centers for Disease Control and Prevention (CDC). https://www.cdc.gov/drugresistance/pdf/threats-report/Actions-For-Healthcare-508.pdf

[10]   Köhler GA, Gong X, Bentink S, *et al.* The functional basis of mycophenolic acid resistance in Candida albicans IMP dehydrogenase. J Biol Chem 2005; 280(12): 11295-302.
[http://dx.doi.org/10.1074/jbc.M409847200] [PMID: 15665003]

[11]   Harrison F, Roberts AEL, Gabrilska R, Rumbaugh KP, Lee C, Diggle SP. A 1,000-year-old antimicrobial remedy with antistaphylococcal activity. MBio 2015; 6(4)e01129-15
[http://dx.doi.org/10.1128/mBio.01129-15] [PMID: 26265721]

[12]   Pallasch TJ. Antibiotics: past, present, and future. CDA J 1986; 14(5): 65-8.
[PMID: 3458545]

[13]   Kitchin JES, Pomeranz MK, Pak G, Washenik K, Shupack JL. Rediscovering mycophenolic acid: A review of its mechanism, side effects, and potential uses. J Am Acad Dermatol 1997; 37(3): 445-9.
[http://dx.doi.org/10.1016/S0190-9622(18)30747-3] [PMID: 9308561]

[14]   Downing HJ, Pirmohamed M, Beresford MW, Smyth RL. Paediatric use of mycophenolate mofetil. Br J Clin Pharmacol 2013; 75(1): 45-59.
[http://dx.doi.org/10.1111/j.1365-2125.2012.04305.x] [PMID: 22519685]

[15]   Baber M. The use and abuse of antibiotics. BJOG. 1960;67 (5): 727-32.
[http://dx.doi.org/10.1111/j.1471-0528.1960.tb10425.x]

[16]   Boucher HW. Bad bugs, no drugs 2002-2020: Progress, challenges, and call to action. Trans Am Clin Climatol Assoc 2020; 131: 65-71.
[PMID: 32675844]

[17]   Boucher HW, Talbot GH, Bradley JS, *et al.* Bad bugs, no drugs: no ESKAPE! An update from the Infectious Diseases Society of America. Clin Infect Dis 2009; 48(1): 1-12.
[http://dx.doi.org/10.1086/595011] [PMID: 19035777]

[18]   WHO. Global Antimicrobial Resistance and Use Surveillance System (GLASS). https://www.who.int/initiatives/glass2023

[19]   Black Z, Balta I, Black L, Naughton PJ, Dooley JSG, Corcionivoschi N. The fate of foodborne pathogens in manure treated soil. Front Microbiol 2021; 12781357
[http://dx.doi.org/10.3389/fmicb.2021.781357] [PMID: 34956145]

[20]   Wimalarathna HML, Richardson JF, Lawson AJ, *et al.* Widespread acquisition of antimicrobial resistance among Campylobacter isolates from UK retail poultry and evidence for clonal expansion of resistant lineages. BMC Microbiol 2013; 13(1): 160.
[http://dx.doi.org/10.1186/1471-2180-13-160] [PMID: 23855904]

[21]   Veltcheva D, Colles FM, Varga M, Maiden MCJ, Bonsall MB. Emerging patterns of fluoroquinolone resistance in *Campylobacter jejuni* in the UK [1998–2018]. Microb Genom 2022; 8(9)mgen000875.
[http://dx.doi.org/10.1099/mgen.0.000875] [PMID: 36155645]

[22]   Hölzel CS, Tetens JL, Schwaiger K. Unraveling the role of vegetables in spreading antimicrobial-

resistant bacteria: A need for quantitative risk assessment. Foodborne Pathog Dis 2018; 15(11): 671-88.
[http://dx.doi.org/10.1089/fpd.2018.2501] [PMID: 30444697]

[23] Initiatives for Addressing Antimicrobial Resistance in the Environment: Current Situation and Challenges. Wellcome Trust. 2018. https://wellcome.ac.uk/sites/default/files/ antimicrobial-resistanc--environment -report.pdf

[24] Pulford CV, Wenner N, Redway ML, *et al.* The diversity, evolution and ecology of Salmonella in venomous snakes. PLoS Negl Trop Dis 2019; 13(6)e0007169.
[http://dx.doi.org/10.1371/journal.pntd.0007169] [PMID: 31163033]

[25] Tapia-Arreola AK, Ruiz-Garcia DA, Rodulfo H, Sharma A, De Donato M. High Frequency of Antibiotic Resistance Genes (ARGs) in the Lerma River Basin, Mexico. Int J Environ Res Public Health 2022; 19(21): 13988.
[http://dx.doi.org/10.3390/ijerph192113988] [PMID: 36360888]

[26] Leonard AFC, Zhang L, Balfour AJ, *et al.* Exposure to and colonisation by antibiotic-resistant *E. coli* in UK coastal water users: Environmental surveillance, exposure assessment, and epidemiological study (Beach Bum Survey). Environ Int 2018; 114: 326-33.
[http://dx.doi.org/10.1016/j.envint.2017.11.003] [PMID: 29343413]

CHAPTER 2

Bacteriophages

Daniel Huelgas-Méndez[1], Juan Carlos García-Cruz[1], José Luis Díaz-Nuñez[2], Xareni Rebollar Juárez[1], Daniela Luis-Yong[3], Mariel Hernández-Garnica[1], Jorge Santiago Jiménez-Zúñiga[1] and Rodolfo García-Contreras[1,*]

[1] *Microbiology and Parasitology Department, Faculty of Medicine, National Autonomous University of México, México City, México*

[2] *Postgraduate in Botany, Comecyt–College of Postgraduates, México*

[3] *Center for Research in Applied Mycology, Veracruzana University, Veracruz, México*

Abstract: Bacteriophages are the most abundant biological entities on the planet and are specific viruses that target only bacteria. The use of these viruses to combat bacterial infections is known as phage therapy, a concept that was implemented in the early 20th century. However, its application in Western medicine was halted following the discovery and use of antibiotics. Still, due to the alarming increase in antibiotic resistance we are experiencing, phage therapy is gaining acceptance in Western countries, and several successful cases have been documented. In this chapter, we discuss various aspects of phage therapy, from bacteriophage isolation and characterization to the development of phage-suitable formulations and their administration for treating human infections. We also examine successful cases of phage therapy in treating life-threatening infections that are untreatable with current antibiotics, highlighting the advantages and limitations of phage therapy compared to traditional antimicrobials.

Keywords: Antibiotic resistance, Bacteriophage, Bacterial infections, Clinical cases, Clinical trials, Persister cells, Phage administration, Phage cocktails, Phage isolation, Biofilms.

INTRODUCTION

Bacteriophages are viruses that infect bacteria and were discovered independently by Frederick Twort and Felix d'Herelle. In 1915, Frederick Twort reported the appearance of lytic plaques at the edge of some *Staphylococcus* colonies and obtained the same pattern when he spotted the filtrate supernatant of the cultures of that bacterium on the lawn of different *Staphylococcus* strains; unfortunately,

---

* **Corresponding author Rodolfo García-Contreras:** Microbiology and Parasitology Department, Faculty of Medicine, National Autonomous University of México, México City, México; E-mail: rgarc@bq.unam.mx

Mariano Martínez-Vázquez (Ed.)

he was unable to explain this phenomenon. Later, in 1917, Felix d'Herelle performed research focused on an outbreak of severe hemorrhagic dysentery among French troops. To get a vaccine, he obtained bacterium-free filtrates of the patients' fecal samples and incubated them with *Shigella* strains isolated from the patients. He observed the appearance of clear lysis plaques. D'Herelle published these results and suggested that the causal agent was a virus capable of infecting bacteria (this was confirmed later by the development of electron microscopy), and he also named them bacteriophages. Immediately after the discovery, the potential of these bacteria to treat human diseases was realized by D'Herelle and other scientists, and hence, they were used successfully for the treatment of dysentery, cholera, and other infections [1]. George Eliava was a Soviet scientist who founded the Eliava Institute of Bacteriophage, Microbiology, and Virology (EIBMV) in 1923. He was aware of D'Herelle's research and invited him to collaborate at his institute. At that time, the institute produced phage preparations against many enteric pathogens, such as *Staphylococcus*, *Pseudomonas*, and *Proteus*. Nowadays, this institute is a worldwide reference for phage cocktail production for various purposes; its collection spans over 1000 phages active against human, plant, and animal bacterial pathogens isolated worldwide. In this chapter, we explain the procedures for isolating and characterizing bacteriophages suitable for phage therapy, including the advantages and limitations they have and how preparations are formulated and administered, as well as discuss several well-documented examples of their utilization for the treatment of recalcitrant bacterial infections in Western medicine.

ISOLATION AND CHARACTERIZATION OF BACTERIOPHAGES SUITABLE FOR PHAGE THERAPY

Bacteriophages are nature's most abundant biological entity, outnumbering bacteria by an order of magnitude. They can be found in any place where bacteria are also found. It is estimated that there are 10^{31} phages in Earth's ocean; meanwhile, there are around 10^{22-24} stars in the observable universe [2]. They are commonly obtained from water sources, such as sewage, contaminated rivers, lakes, and the sea, but can also be obtained from soil, feces, saliva, sputum, and other secretions [3 - 6]. They are intracellular parasites that rely on a bacterial host to replicate their genome and assemble into mature forms called virions. The infection begins with the viral recognition of phage receptors located on the bacterial surface, and then the phage injects its genome into its bacterial host. Once the genome enters the cell, the phage can adopt two life cycles: a lytic or a lysogenic cycle. During the lytic cycle, the phage genome starts its replication, transcription, and translation. The viral particles then assemble, and finally, the bacterial host is lysed. In the lysogenic cycle, the phage genome integrates into the bacterial DNA to form a non-infective form called a prophage, which is replicated along with the bacterial genome. Nevertheless, the prophages can excise the bacterial genome stochastically or upon exposure to certain types of stress and switch to a lytic life cycle [7]. For their application in phage therapy, lytic phages are strongly preferred over temperate ones since their life cycle involves invading and killing their host. In contrast, although temperate phages can undergo lytic cycles, they typically remain dormant most of the time. At the same time, their genomes are replicated each time their host replicates, becoming active or lytic under certain environmental conditions, often triggered by factors such as oxidative stress or quorum sensing [8].

However, since most phages are temperate, they could be engineered to produce obligate lytic versions by removing genes essential for the lysogenic cycle, such as repressors and integrases. This approach could enhance phage therapy, thereby increasing the repertoire of phages available for clinical use.

An established technology in phage engineering is phage display, which involves using viroid particles or capsid proteins to deliver substances, drugs, or plasmids, taking advantage of host specificity. This process leads to the creation of phagemids. It has been studied for inducing the expression of antimicrobial peptides that signal an immune response (immunization/vaccination), as well as for anticancer therapy, and other applications [10].

Other vital points that must be considered are that it should be avoided utilizing phages that contain potentially dangerous genes, such as those encoding virulence factors or antibiotic resistance determinants, since they may increase the resistance and/or virulence of their host and other bacteria. Hence, it is advisable to sequence the phage's genomes before their utilization and either exclude those that carry dangerous genes or engineer them to remove those genes that encode harmful components. Furthermore, phages should be thoroughly purified prior to their administration to avoid the delivery of toxic bacterial components like LPS that could cause septic shock [9].

An important feature of phage therapy is that it should be implemented not with single phages, but rather with a combination of several different ones, since single phages can easily select for bacterial resistance. The probability that bacteria with resistance against multiple phages will emerge is very low if the phages used are different, for example, if they do not share the same receptor [9].

Although, in principle, phage therapy could be applied to any pathogenic bacteria, to develop robust phage therapies, it is necessary to have extensive collections of different phages targeting the bacterial host; the optimal size of the collection is variable, since for some bacteria like *Staphylococcus aureus* phages usually can infect several different strains, in contrast for other bacterial pathogens, such as *Acinetobacter baumannii*, phages have a narrow host range, sometimes being specific for one or a few strains, hence is estimated that for *S. aureus* an optimal collection should contain at least 20 phages, while for *A. baumannii* it should contain around 300 [9, 11].

ADVANTAGES OF PHAGE THERAPY OVER CLASSICAL ANTIBACTERIAL AGENTS

Unlike classical antimicrobials, bacteriophages are biological entities with the ability to self-replicate at the expense of bacteria; hence, their dynamics inside a host are very different from that of antibiotics since chemicals are metabolized or degraded in the liver and eliminated *via* urine. In contrast, as long as phages can reach the target bacterial population, their concentration will increase depending on the amount of susceptible hosts available and decrease when it host population decreases, hence self-adjusting to their target. Also, in general, most of the phages characterized to date are specific to a bacterial species or even to a few or one strain. Hence, it is expected that they may have fewer side effects than regular broad-spectrum antibiotics that damage several bacterial species, including members of the beneficial microbiome. However, if needed, for example, to treat mixed bacterial infections, broad host phages also exist and can be isolated from the environment [12].

Additionally, although bacteria can readily develop resistance to bacteriophages, as seen with antibiotics, this often introduces trade-offs that may reduce their virulence and/or resistance to antibiotics. This frequently occurs because, in developing resistance to phages, bacteria may lose the expression and/or function of their phage receptors, which include virulence factors such as flagella, type IV pili, capsular components, and even proteins associated with antibiotic efflux pumps. Consequently, this produces favorable side effects, making bacteria less aggressive and easier to treat [13 - 15].

Moreover, when bacteria become resistant to an antibiotic, a dead end is reached, while if bacteria become resistant to the phage, it can also, in principle, evolve and regain the ability to infect the previously resistant host [16], something that, in nature, can be considered a red queen race.

Additional advantages of using phages are that some of them can kill persister cells—a fraction of the bacterial population that is dormant and thus impervious to antibiotics that target metabolic active processes such as cell wall replication, protein synthesis, and nucleic acid replication. Since these antibiotics are ineffective against persisters by definition, phages offer an alternative as they can penetrate the persister cells and kill them after they become metabolically active

[17]. Furthermore, recent evidence suggests that phages can also bind to spores produced by the Gram-positive bacterium *Paenibacillus larvae*, indicating that they may also be able to kill spore-making bacteria upon their activation [18]. Fig. (**1**) shows the advantages of bacteriophages over antibiotics.

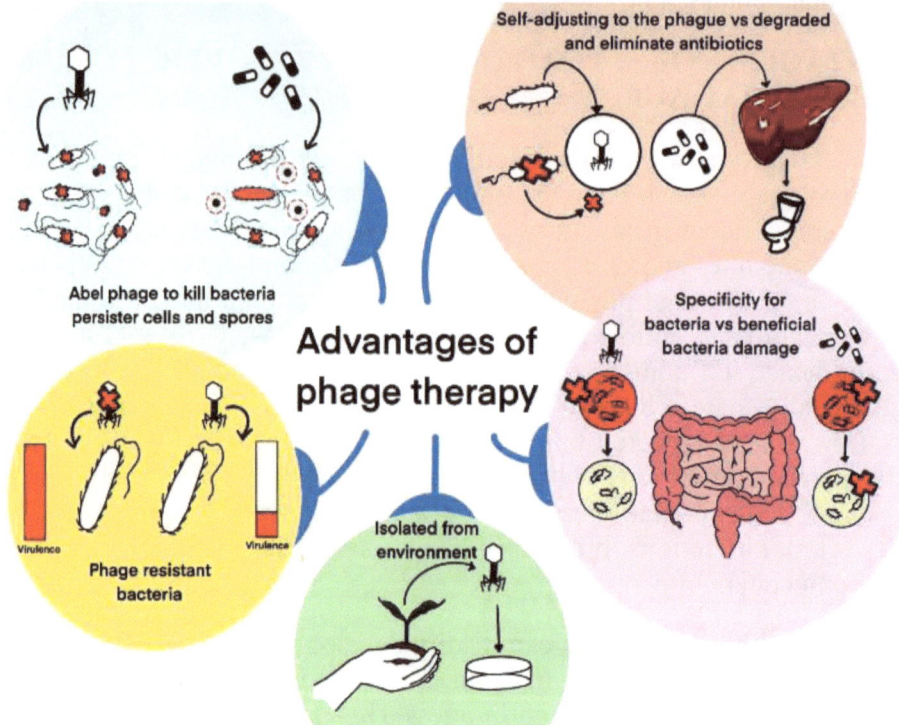

Fig. (1). Advantages of phage therapy.

PHAGE COCKTAILS FORMULATION

The use of phages as therapeutic agents against bacteria that produce a clinically relevant infection is known as phage therapy and can be classified depending on the number of phages used in the preparations in monophage or poliphage therapy [19].

Preparations used for monophage therapy involve only a single phage, and they are mainly used in research during the design or testing of phage preparations. The rapid emergence of phage-resistant clones within the bacterial population is the main reason for the low utility of these preparations [20].

Poliphage therapy counterpart involves using a phage cocktail containing two or more phage types. The use of phage cocktails enables the use of multiple phages against one specific bacterium to decrease the emergence of resistant clones and increase the host range. Additionally, it involves the use of phages that infect different bacteria, which may be the most common etiological agents of a specific disease [19]. Although the use of phage cocktails for human therapy is allowed in some countries, such as Georgia, the Food and Drug Administration (FDA) has not yet approved it for its regular use and only allows it for compassionate use. Nevertheless, some phage preparations like the ones listed in Table **1** have been approved for use in the food industry.

Table **2** shows examples of phage cocktails used to treat human infections in clinical trials and cases.

Table 1. Some FDA-approved phage cocktails. Adapted from a previous study by Jamal M [21].

Product	Company	Target organism(s)	Applications
ListShield™	Intralytix, Inc. (USA)	*Listeria monocytogenes*	Used in various ready-to-eat foods
EcoShield™		Shiga Toxin-producing *Escherichia Coli* (STEC)	Used in food, including meat, poultry, fish, etc, as well as dairy products and vegetables
SalmoFresh™		*Salmonella* enterica	Used mainly in red meat and poultry before grinding
ShigaShield™		*Shigella spp.*	Used in food, including meat, poultry, fish, etc., as well as dairy products and vegetables
PhageGuard Listex	Micreos FoodSafety (The Netherlands)	*Listeria monocytogenes*	Used in various ready-to-eat foods
PhageGuard S		*Salmonella spp.*	It can be used in ground meat and poultry products
PhageGuard E		*E. coli 0157*	It can be used in leafy green vegetables, beef carcass, parts, and trim

PHAGE ADMINISTRATION

Since phage therapies have shown promise in counteracting antibiotic-resistant pathogenic bacterial infections [26 - 28], the search to optimize their administration has become one of the most critical tasks in both clinical and preclinical research [29]. In historical events, it is worth noting that oral and intravenous phage therapies were used to counteract cholera and typhoid fever [30]. Currently, the main forms of phage administration are oral (absorption), injection (intravenous, intramuscular, intraperitoneal, and subcutaneous), inhalation (intranasal and endotracheal), intrarectal (absorption), and topical [29, 31] (Table **3**).

Table 2. Recent phage cocktails used to treat human infections.

Bacteria	Host	Characteristic	Administration	Phages	Results	Comments	Ref.
Enterococcus spp, *Escherichia coli, Proteus mirabilis, Pseudomonas aeruginosa, Staphylococcus* spp, and *Streptococcus* spp.	Men with UTIs and catheter-associated UTIs	Double-blind clinical trial	Pyophage; intravesical twice daily for 7 days	Pyophage cocktail	Intravesical bacteriophage therapy was the same as antibiotic treatment	A dose has not yet been established, but it is not harmful	[22]
Pseudomonas aeruginosa	Patients who had a burn wound infected with *P. aeruginosa*	Randomized, controlled, double-blind phase 1/2 trial	12 lytic phages anti-*P aeruginosa*: daily topical treatment for 7 days, and follow-up for 14 days	PhagoBurn	More than 50% of the participants improved	It is independent of the use of systemic ATB with the strain	[23]
Mycobacterium abscessus	A 15-year-old cystic fibrosis patient with a disseminated *Mycobacterium abscessus* infection	Clinical case	5 mL unit dose IV twice daily. Then the patient continued at home with a central catheter for an additional nine days	A three-phage anti-*M. abscessus* GD01 cocktail	Efficiently killed the infectious *M. abscessus* strain	Good tolerance	[24]
Cutibacterium acnes	Sowing of skin extracts from mild-to-moderate acne vulgaris patients	Phase 1 randomized clinical trial	10 µL of BX001 in a gel, with different inserts. At 24 and 48 hours	Three-phage cocktail, BX001	24% reduction in *C. acnes* levels	Demonstrated that are active versus bacteria embedded in biofilm	[25]

Table 3. Routes of Bacteriophage Administration, Adapted from previous studies [29, 31].

Routes	Penetration into the blood	Infection
Oral and Intrarectal*	Very weak (41.1%)	Typhoid fever, Dysentery, Cholera, Bacterial, Diarrhea, and Prostatitis*
Injection	Very good (98.5%)	Systemic infections
Inhalation	Medium (66.7%)	Tuberculosis
Topical	Weak (50%)	Chronic wounds Burns wounds

The oral administration of phages is the most attractive alternative because it is simple, feasible, and efficient [29, 32]. The efficiency of the route for phages to reach the blood is low (41.1%), and their dispersion in organs and tissues is moderate, where they can be affected by the pH and digestive enzymes of the stomach and gastrointestinal tract, as well as liver bile [31]. Furthermore, although oral phage formulations are suitable, the main problem is that their availability and dosing kinetics are unknown [5, 33, 34].

Attempts are currently being made to approve an oral therapy with phages that began in 2021 and is in clinical phase 1 [35]. LMN-201 is a phage-derived endolysin and three toxin-binding proteins that are intended to be used as a therapy to treat infections caused by *C. difficile* [35].

In the case of administration by injection (mainly intravenous), the feasibility of the route for phages to reach the blood is high (98.5%), and their dispersal in organs and tissues can be good or moderate, depending on the organ [31]. The main problem of this approach is the production of phage combinations that are free from bacterial contaminants [29].

Attempts have recently been made to approve four therapies using intravenously injected phages, which began in 2020 and are in phases 1/2, 1, 2a/1, and 3 clinical trials [35]. In phase 1/2 we have personalized phage therapies in a study called "PhageBank (process)"; in phase 1 is a phage cocktail treatment (CRISPR technology), which will be used to treat infections caused by *E. coli* and in phase 2a/1 and 3 are the phage combinations, LSVT-1701 (N-Rephasin SAL200, tonabacase) and Exebacase (CF-301), that will be used to treat infections caused by *S. aureus* [36].

The administration of phages by inhalation is thought to be easy [28, 37], their assimilation rate in blood is medium (66.7%), and their dispersion in organs and tissues can be excellent. In addition, bacterial contamination may have little impact on the penetration of phages through the respiratory tract [31]. However, the preparation of the treatments is a financial and logistical challenge [29]. In the case of inhaled treatments, they are trying to approve four that are in phase 1/2 and 3 of clinical trials [35]. In phase ½, the phage therapies AP-PA02, YPT-01, and BX004-A are sought to be applied to treat infections caused by *P. aeruginosa*. While in phase 3, a phage formulation is being developed against tonsillitis in children [35]. Topical and intrarectal administration of phages may be an effective and feasible alternative [29, 31]. However, the possibility of intrarectal therapies with phages has been little studied [22], and in the case of topical application, the skin can act as a barrier that prevents the penetration of phages through their capsids [31, 38].

SUCCESS OF PHAGE THERAPY AGAINST MULTI-DRUG-RESISTANT AND PAN-RESISTANT BACTERIA

Drug-resistant infections are a global issue with health and economic dimensions. It has been estimated by the economist Jim O'Neill that if the problem is tackled with any public policy, the number of estimated annual deaths by 2050 would be 10 million, which means one person every three seconds, and 100 trillion USD in terms of lost global production. This trend is on the rise, particularly in low and middle-income countries. Indeed, a recent study has estimated that in 2019, 4.5 million deaths were associated with bacterial AMR, and 1.2 million deaths were attributed to the same problem [39, 40].

Moreover, developing and marketing new antibiotics pose challenges, such as optimizing systemic therapy in humans and narrowing the market niche. That is why it has been recommended to promote the use and development of antibiotic alternatives, such as wild-type or engineered bacteriophages.

The World Health Organization (WHO) has listed some bacteria according to their urgency in developing new antibiotics due to their increasing antibiotic resistance. In the list of "critical urgency" members of the ESKAPE group are included: *Acinetobacter baumanii* carbapenem resistant, *Pseudomonas aeruginosa* carbapenem resistant, and *Enterobacteriaceae* carbapenem and 3rd generation cephalosporin resistant. Here, some of the cases of success in the use of phage therapy against multi-drug and pan-drug-resistant bacteria will be discussed.

A case from a 68-year-old patient with diabetes named Tom Patterson has been well covered by the media in the US and has brought back the interest of phage therapy in the West. The case is narrated in the first person by him and his wife, Steffanie Strathdee, in the book The Perfect Predator.

The couple went to Egypt for Thanksgiving vacation when Tom started to feel stomach and back pain. The doctors at the hospital in Egypt diagnosed acute pancreatitis, but due to the lack of proper equipment to attend to Tom's illness, they transferred them to Germany, where Tom was diagnosed with necrotizing acute pancreatitis and a complication of a pseudocyst in the abdomen. The physicians also isolated an MDR *A. baumannii* and a fungus, *Candida glabrata,* from a sample of the pseudocyst fluid. Next, both Tom and Steffanie moved back to their home in San Diego, California, where Tom was hospitalized at the UCSD hospital, a medical center experienced in treating *A. baumannii* infections. The prognosis was not very optimistic because the bacteria had already become resistant to all available antibiotics.

As Tom was in a coma, Steffanie started looking for alternative therapies and found the paper [41]. So, she sent emails to several phage research groups in the US and abroad and received a quick response from Ry Young, who had been working with phages for almost 45 years at the Center for Phage Technology at Texas A&M University. In parallel, their doctor and close friend, Robert Schooley, was in dialogue with the FDA to know how to manage the phage therapy for Tom's case, and they suggested administering it through the emergency Investigational New Drug (IND) framework, which was the first time FDA-approved phage therapy was done through this pathway. Also, they informed the US Naval Medical Research Center, which had been working on phages against MDR *A. baumannii* infections.

The group of Texas found four phages that efficiently infected Tom's isolate; the only problem was that the endotoxin levels were too high, so they needed to send the phage cocktail to a local academic group in San Diego for additional endotoxin removal [42, 43]. Before Tom was dialyzed, the phages arrived at the hospital and were administered through percutaneous catheters draining the pseudocyst cavity, the biliary cavity, and a third intra-abdominal cavity [43]. The therapy had no adverse effects, so the phages were administered every 2 hours. After 4 days, the US Navy phage cocktail arrived and was administered intravenously. Two days later, Tom awakened from his deep coma. As far as is known, he became the first US patient to be treated with phage therapy for a multisystemic MDR infection.

The phage therapy lasted for almost two months, along with Meropenem, Minocycline, and Fluconazole, until the *A. baumannii* levels reached a point at which Tom's immune system was able to manage. He had some sequels like insulin-dependent diabetes, mild heart damage, not feeling in the bottom of his feet, and gut damage, but he, his wife, and all the scientists and health workers involved in the case impulse initiatives in the US like the Center for Innovative Phage Applications and Therapeutics (IPATH) directed by the Dr. Schooley who has already started clinical trials for phages against MDR *P. aeruginosa* in cystic fibrosis cases. Moreover, the US Navy began to work with a biotechnology company called Adaptive Phage Therapeutics, which has also helped in other phage therapy for compassionate use cases and has a big biobank of phages against different MDR bacteria.

Another recent successful case of phage therapy involved a young male from Algeria who had biliary atresia [44]. He was transferred to the Saint Luc University Hospital in Belgium and waited for liver transplantation. Twenty days after the transplant, the child acquired an infection of XDR *P. aeruginosa* that was only sensitive to colistin and had intermediate resistance to aztreonam.

After 53 days, he had multiple hepatic abscesses and severe septicemia caused by the *P. aeruginosa* strain. Since the intravenous administration did not cure his infection with a cocktail of antibiotics, including colistin, gentamicin, and aztreonam, phage therapy was initiated. The phage cocktail "BFC1" was produced by the Queen Astrid Military Hospital and contained one *S. aureus* phage and two phages against *P. aeruginosa*. Before clinical application, the cocktail was analyzed for approval by the Belgian Institute for Public Health, with some of the criteria including the absence of any genetic determinant coding for lysogeny, toxins, or antibiotic resistance, evidence that the phages were lytic, and that the preparations had low endotoxin levels.

After BFC1 approval, 1mL/kg of body weight was administered daily via a 6-hour IV infusion. Thirty-six hours after phage therapy, two subsequent blood cultures tested negative for bacteria. However, between days 4 and 7, after the beginning of phage therapy, blood and pus abscess cultures were positive for *P. aeruginosa*. So, on day 8, the IV phage dose was doubled (2 mL/kg). This measure allows the maintenance of negative blood cultures for *P. aeruginosa*. However, two transient translocations of this bacterium from the liver abscesses to the bloodstream were detected on days 27 and 30. They were controlled in less than 24 hours without any new therapeutic intervention.

Phage therapy in combination with antibiotics eliminated the *P. aeruginosa* infection in the bloodstream but not in the hepatic lesions. The patient showed a stable state outside the intensive care unit with intermittent fever and persistent inflammation and waited until a postmortem-compatible liver was available for retransplantation after 3 days of phage therapy. During the liver transplantation, the abdominal cavity was washed with 250 mL of BCF1. In total, the patient received 86 uninterrupted days of IV phage therapy, and no adverse effect was observed concurrently, with a total clearance of his *P. aeruginosa* infection.

In 2016, a 30-year-old Belgian woman suffered from a bomb attack in Brussels and was admitted to the intensive care unit of the Erasme Hospital; after 4 days of hospitalization, she presented septic shock due to a surgical wound infection of the left thigh. Bacterial cultures revealed a polymicrobial infection consisting of *Enterococcus faecium, P. aeruginosa, Enterobacter cloacae,* and *Klebsiella pneumoniae*. So, a high-dose, broad-spectrum, long-term antibiotic course was given, leading to several adverse effects such as febrile neutropenia and post-traumatic deafness.

After 170 days, surgical biopsies from the femur showed the presence of 2 *K. pneumoniae* strains. One has an XDR phenotype, being only sensitive to amikacin and fosfomycin. So, an adapted antibiotic regimen was started. This regimen failed to clear the infection, and hence, phage therapy was initiated. Both isolates of *K. pneumoniae* were sent to the Eliava Institute to select and adapt therapeutically beneficial phages. Phage vB_KpnM_M1 (M1) presented the highest activity against the patient's isolates. Remarkably, this phage had a broad host range (~65%) against clinical isolates of different species of *Klebsiella.*

Furthermore, an evolutionary approach was used to "train" this phage against the patient's isolates, co-evolving them with the phage M1 in liquid media [45]. The result was improved lytic activity after fifteen rounds of co-evolution. In addition, the phage genome lacked genes related to lysogeny, toxins, or antibiotic resistance.

The hospital's ethical committee approved using this phage for therapy in XDR *K. pneumoniae* Fracture-Related Infection (FRI) in November 2016. However, it was put on hold due to a lack of consensus among the physicians. On 21 February, 2018, facing a therapeutic dead end, the phage preparation was applied at the end of a surgical intervention; the phages were administered locally to obtain the highest concentration at the site of infection. In addition, a pan-drug-resistant *K. pneumoniae* isolate was confirmed from a biopsy taken during surgery. The phages were administered for 6 days (days 702 to 707 post-injury) to minimize possible immunogenicity, and the ongoing antibiotic treatment continued until day 798 post-injury.

No adverse events related to the use of these phages were observed. Comparing the patient's clinical, biological, microbiological, and radiological conditions before and after the combined antibiotic-phage therapy intervention, it was concluded that the infection was controlled, and the patient's general condition improved. By 11 June, 2018 (day 806 post-injury and 104 posts antibiotic-phage therapy), there was no growth of *K. pneumoniae*, and the external fixator from the patient's hip and leg was removed, considered clinically cured. Table (**4**) summarizes successful cases of phage therapy for treating human infections.

Table 4. Cases of successful phage therapy against multi-drug and pan-resistant bacteria.

Multi-/Pan- drug resistant pathogen	Infection	Route of administration	Phage origin	Duration	Ref.
A. baumannii	Necrotizing pancreatitis	IV, local	Environmental, military, phage bank, biotech	~ 2 months	[43]
P. aeruginosa	Hepatic abscesses; severe septicemia	IV, local	Military	~ 3 months	[46]

(Table 4) cont.....

Multi-/Pan- drug resistant pathogen	Infection	Route of administration	Phage origin	Duration	Ref.
K. pneumoniae	Explosion wound on left flank and thigh	Local	Environmental	6 days	[47]
A. baumannii, K. pneumoniae	Bone	IV	Military	11 days	[48]
P. aeruginosa	Recurrent pneumonia	Inhaled; IV	Environmental, military, biotech	~2.5 months	[49]
P. aeruginosa	Catheter-related bacteremia	IV, local	Military, biotech	21 days	[50]
S. aureus	Corneal abscess	Topical, nasal, IV	Commercial (Eliava Institute)	4 weeks	[51]

LIMITATIONS OF PHAGE THERAPY

Phage therapy has existed for many years but has only gained prominence in occidental medicine in the last 20 years. The first trials of phage therapy piloted by d'Herelle were relevant but poorly controlled, for example, lacking a placebo group and focusing only on administering phages to sick patients, believing that this therapy could save the lives of ill people [52]. Currently, phase 3 clinical trials for phage therapy that include many patients are still difficult to conduct since it is challenging to find patients infected with the same strain or with strains susceptible to the same phage, given that this therapy is a kind of personalized medicine, rather than a generic standard procedure.

By the beginning of the 20th century, phage preparations were marketed in Brazil and the United States. Still, the results were unfavorable, as there was a preference for antibiotics. The limitations of the early trials occurred since, at that time, the knowledge about the biological nature of phages was scarce [53], resulting in a low rate of active phages and contamination of phage preparations, so phage therapy in the 1940s was discarded.

On the other hand, the American Medical Association ordered a thorough review to consider the viability of phage therapy, which unfortunately resulted in unfavorable conclusions; they affirmed that "the theory put forward by d'Herelle had not been proved" [54]. It is now known that many of the hypotheses put forward at the time were incorrect. Still, the research deficiency at the time dealt a severe negative blow to several corporations willing to fund therapeutic phage research.

The effectiveness of bacteriophage therapy remains uncertain, even though all the properties of lytic phages have been demonstrated. They are not yet widely used as therapeutics in Western medicine, and there are still doubts about their feasibility.

There are adverse reactions in the oral administration of phages; these suffer a translocation when they reach the intestinal epithelium on some occasions. This turns out to be beneficial for the host because they regulate the low immune response to the antigens of the intestinal bacterial flora, and this advantage is promising [55]. There is evidence that bacteriophages kill bacteria in their immunocompromised host [56]. Additionally, other authors describe the possibility of an adverse reaction in immunocompromised patients.

Another limitation is the possibility that the use of phage cocktails may affect the state of the intestinal barrier to what is known as "leaky gut" to test if the effect was real [57], experimented in an animal model to demonstrate whether oral phage administration affected intestinal permeability by elevating serum levels of inflammatory surrounding immunocompromised in the blood associated with pathological conditions, and showed that oral phage therapy would be associated with intestinal barrier disorders such as Crohn's disease and type 1 diabetes [57].

Specific points should be considered before choosing this therapy as antibacterial, phages should be "lytic" with a high potential to resist different conditions [58], but lytic phages are not exempt from concerns their genomes may contain more than 50% of genes with unknown function [59] they may encode proteins that alter bacterial physiology, the promise of phage therapy requires characterization of their genomes to evaluate candidate phages. Although phages are host-specific, they can switch hosts in the gut. This allows them to adapt and evolve, giving them a significant advantage but a minor disadvantage when applied in therapy [60]. Phage therapy alone is unlikely to eliminate an infection, and eradicating the host will result in the termination of viral replication [20]. The efficiency of treatment depends on the timing of phage dosing. In one study, the *Klebsiella pneumoniae*-specific SS phage B5055 was administered intranasally to mice. Dosing the phage three hours before bacterial infection significantly protected the mice, and dosing six hours after treatment was unsuccessful [61]. The authors suggest that timing is vital for success. Phage cocktail preparations require a long preparation time and increase the likelihood of provoking immune responses, which can affect the phages' pharmacokinetic and pharmacodynamic properties [19].

Moreover, some host components, such as mucin in the intestine, can bind and decrease phage infectivity. Nevertheless, some phages are impervious to this inhibitory effect and even show enhanced activity in the gastrointestinal tract. Furthermore, the host sometimes produces neutralizing antibodies against administered phages, neutralizing their therapeutic effect.

Not all phages are suitable candidates for their therapeutic use; they must have high potential, be stable lytic, resist temperature change, and be correctly identified. The phages should be well characterized to avoid those with low potential against bacterial targets, and the use of phages with poor pharmacokinetics should be avoided. Phage therapy should be limited to treating infections when antibiotics are ineffective [62]. It should be noted that phage replication may increase the release of endotoxins, which are inducers of the inflammatory response of cytokinins in Gram-negative bacteria [63]. These could also be present in phage preparations. Endotoxin is challenging to separate from the phage and induces symptoms that include redness, heart rate disturbances, blood pressure imbalance, dyspnea, bronchospasm, back pain, fever, urticaria, edema, nausea, and rash [64]. The data reported by Zhang *et al.* [65] from studies carried out *in vitro* and *in vivo* by Park *et al.* [66] described that phages directly affect innate and adaptive immunity [44]. Fig. (**2**) summarizes the current limitations of phage therapy.

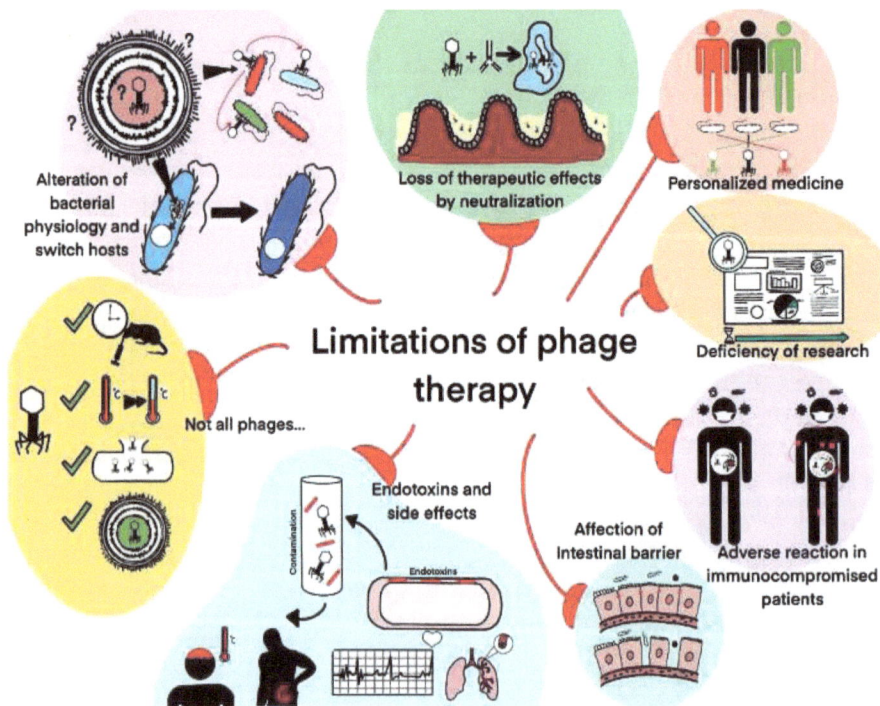

Fig. (2). Limitations of phage therapy.

CONCLUSION

Although bacteriophages represent a viable alternative to antibiotics to combat recalcitrant bacterial infections, their establishment as part of mainstream Western medicine is still far off due to limitations precluding their systematic utilization in many cases. Nevertheless, it is anticipated that, due to the current antibiotic resistance crisis, they will gradually be adopted as therapeutic tools to treat a wide variety of infections, and will likely be used widely in the future as part of a personalized medicine scheme.

ACKNOWLEDGEMENTS

RG-C research is supported by PAPIIT-UNAM Grant IN200224. Díaz-Nuñez J.L. was supported by Cátedra- COMECYT program (ESYCA2023-117253). DH-M was supported by the Programa de Maestría y Doctorado en Ciencias Bioquímicas, UNAM and a scholarship for doctoral studies by SECIHTI, Mexico (CVU number: 1103451).

REFERENCES

[1] Chanishvili N. Phage therapy-history from Twort and d'Herelle through Soviet experience to current approaches.Adv Virus Res. 2012; 83: pp. 3-40.
[http://dx.doi.org/10.1016/B978-0-12-394438-2.00001-3]

[2] Grose JH, Casjens SR. Understanding the enormous diversity of bacteriophages: The tailed phages that infect the bacterial family Enterobacteriaceae. Virology 2014; 468-470: 421-43.
[http://dx.doi.org/10.1016/j.virol.2014.08.024] [PMID: 25240328]

[3] Cross T, Schoff C, Chudoff D, *et al.* An optimized enrichment technique for the isolation of Arthrobacter bacteriophage species from soil sample isolates. J Vis Exp 2015; 52781(98): 52781.
[http://dx.doi.org/10.3791/52781] [PMID: 25938576]

[4] Bachrach G, Leizerovici-Zigmond M, Zlotkin A, Naor R, Steinberg D. Bacteriophage isolation from human saliva. Lett Appl Microbiol 2003; 36(1): 50-3.
[http://dx.doi.org/10.1046/j.1472-765X.2003.01262.x] [PMID: 12485342]

[5] Chibani-Chennoufi S, Sidoti J, Bruttin A, *et al.* Isolation of *Escherichia coli* bacteriophages from the stool of pediatric diarrhea patients in Bangladesh. J Bacteriol 2004; 186(24): 8287-94.
[http://dx.doi.org/10.1128/JB.186.24.8287-8294.2004] [PMID: 15576777]

[6] Shende RK, Hirpurkar SD, Sannat C, Rawat N, Pandey V. Isolation and characterization of bacteriophages with lytic activity against common bacterial pathogens. Vet World 2017; 10(8): 973-8.
[http://dx.doi.org/10.14202/vetworld.2017.973-978] [PMID: 28919692]

[7] Campbell A. The future of bacteriophage biology. Nat Rev Genet 2003; 4(6): 471-7.
[http://dx.doi.org/10.1038/nrg1089] [PMID: 12776216]

[8] Hu J, Ye H, Wang S, Wang J, Han D. Prophage activation in the intestine: Insights into functions and possible applications. Front Microbiol 2021; 12785634.
[http://dx.doi.org/10.3389/fmicb.2021.785634] [PMID: 34966370]

[9] Strathdee SA, Hatfull GF, Mutalik VK, Schooley RT. Phage therapy: From biological mechanisms to future directions. Cell 2023; 186(1): 17-31.
[http://dx.doi.org/10.1016/j.cell.2022.11.017] [PMID: 36608652]

[10]   Hess KL, Jewell CM. Phage display as a tool for vaccine and immunotherapy development. Bioeng Transl Med 2020; 5(1)e10142.
       [http://dx.doi.org/10.1002/btm2.10142] [PMID: 31989033]

[11]   Göller PC, Elsener T, Lorgé D, *et al.* Multi-species host range of staphylococcal phages isolated from wastewater. Nat Commun 2021; 12(1): 6965.
       [http://dx.doi.org/10.1038/s41467-021-27037-6] [PMID: 34845206]

[12]   Cazares D, Cazares A, Figueroa W, Guarneros G, Edwards RA, Vinuesa P. A novel group of promiscuous podophages infecting diverse gammaproteobacteria from river communities exhibits dynamic intergenus host adaptation. mSystems 2021; 6(1)e00773-20.
       [http://dx.doi.org/10.1128/mSystems.00773-20] [PMID: 33531404]

[13]   Kortright KE, Done RE, Chan BK, Souza V, Turner PE. Selection for phage resistance reduces virulence of *Shigella flexneri*. Appl Environ Microbiol 2022; 88(2)e01514-21.
       [http://dx.doi.org/10.1128/AEM.01514-21] [PMID: 34788068]

[14]   Chan BK, Sistrom M, Wertz JE, Kortright KE, Narayan D, Turner PE. Phage selection restores antibiotic sensitivity in MDR *Pseudomonas aeruginosa*. Sci Rep 2016; 6(1): 26717.
       [http://dx.doi.org/10.1038/srep26717] [PMID: 27225966]

[15]   Gordillo Altamirano F, Forsyth JH, Patwa R, *et al.* Bacteriophage-resistant Acinetobacter baumannii are resensitized to antimicrobials. Nat Microbiol 2021; 6(2): 157-61.
       [http://dx.doi.org/10.1038/s41564-020-00830-7] [PMID: 33432151]

[16]   Castledine M, Sierocinski P, Inglis M, *et al.* Greater phage genotypic diversity constrains arms-race coevolution. Front Cell Infect Microbiol 2022; 12834406
       [http://dx.doi.org/10.3389/fcimb.2022.834406] [PMID: 35310856]

[17]   Tkhilaishvili T, Lombardi L, Klatt AB, Trampuz A, Di Luca M. Bacteriophage Sb-1 enhances antibiotic activity against biofilm, degrades exopolysaccharide matrix and targets persisters of *Staphylococcus aureus*. Int J Antimicrob Agents 2018; 52(6): 842-53.
       [http://dx.doi.org/10.1016/j.ijantimicag.2018.09.006] [PMID: 30236955]

[18]   Brady TS, Roll CR, Walker JK, *et al.* Phages Bind to Vegetative and Spore Forms of *Paenibacillus larvae* and to Vegetative *Brevibacillus laterosporus*. Front Microbiol 2021; 12588035
       [http://dx.doi.org/10.3389/fmicb.2021.588035] [PMID: 33574806]

[19]   Chan BK, Abedon ST, Loc-Carrillo C. Phage cocktails and the future of phage therapy. Future Microbiol 2013; 8(6): 769-83.
       [http://dx.doi.org/10.2217/fmb.13.47] [PMID: 23701332]

[20]   Gordillo Altamirano FL, Barr JJ. Phage Therapy in the Postantibiotic Era. Clin Microbiol Rev 2019; 32(2)e00066-18
       [http://dx.doi.org/10.1128/CMR.00066-18] [PMID: 30651225]

[21]   Jamal M, Bukhari SMAUS, Andleeb S, *et al.* Bacteriophages: an overview of the control strategies against multiple bacterial infections in different fields. J Basic Microbiol 2019; 59(2): 123-33.
       [http://dx.doi.org/10.1002/jobm.201800412] [PMID: 30485461]

[22]   Leitner L, Ujmajuridze A, Chanishvili N, *et al.* Intravesical bacteriophages for treating urinary tract infections in patients undergoing transurethral resection of the prostate: a randomised, placebo-controlled, double-blind clinical trial. Lancet Infect Dis 2021; 21(3): 427-36.
       [http://dx.doi.org/10.1016/S1473-3099(20)30330-3] [PMID: 32949500]

[23]   Jault P, Leclerc T, Jennes S, *et al.* Efficacy and tolerability of a cocktail of bacteriophages to treat burn wounds infected by *Pseudomonas aeruginosa* (PhagoBurn): a randomised, controlled, double-blind phase 1/2 trial. Lancet Infect Dis 2019; 19(1): 35-45.
       [http://dx.doi.org/10.1016/S1473-3099(18)30482-1] [PMID: 30292481]

[24]   Dedrick RM, Guerrero-Bustamante CA, Garlena RA, *et al.* Engineered bacteriophages for treatment of a patient with a disseminated drug-resistant Mycobacterium abscessus. Nat Med 2019; 25(5): 730-3.

[http://dx.doi.org/10.1038/s41591-019-0437-z] [PMID: 31068712]

[25] Golembo M, Puttagunta S, Rappo U, *et al.* Development of a topical bacteriophage gel targeting *Cutibacterium acnes* for acne prone skin and results of a phase 1 cosmetic randomized clinical trial. Skin Health Dis 2022; 2(2)ski2.93
[http://dx.doi.org/10.1002/ski2.93] [PMID: 35677920]

[26] Cano EJ, Caflisch KM, Bollyky PL, *et al.* Phage Therapy for Limb-threatening Prosthetic Knee *Klebsiella pneumoniae* Infection: Case Report and *In Vitro* Characterization of Anti-biofilm Activity. Clin Infect Dis 2021; 73(1): e144-51.
[http://dx.doi.org/10.1093/cid/ciaa705] [PMID: 32699879]

[27] Aslam S, Lampley E, Wooten D, *et al.* Lessons learned from the first 10 consecutive cases of intravenous bacteriophage therapy to treat multidrug-resistant bacterial infections at a single center in the United States. Open Forum Infect Dis 2020; 7(9)ofaa389
[http://dx.doi.org/10.1093/ofid/ofaa389] [PMID: 33005701]

[28] Maddocks S, Fabijan AP, Ho J, *et al.* Bacteriophage therapy of ventilator-associated pneumonia and empyema caused by *Pseudomonas aeruginosa.* Am J Respir Crit Care Med 2019; 200(9): 1179-81.
[http://dx.doi.org/10.1164/rccm.201904-0839LE] [PMID: 31437402]

[29] Khalid A, Lin RCY, Iredell JR. A Phage Therapy Guide for Clinicians and Basic Scientists: Background and Highlighting Applications for Developing Countries. Front Microbiol 2021; 11599906.
[http://dx.doi.org/10.3389/fmicb.2020.599906] [PMID: 33643225]

[30] d'Herelle F. Bacteriophage as a Treatment in Acute Medical and Surgical Infections. Bull N Y Acad Med 1931; 7(5): 329-48.
[PMID: 19311785]

[31] Dąbrowska K. Phage therapy: What factors shape phage pharmacokinetics and bioavailability? Systematic and critical review. Med Res Rev 2019; 39(5): 2000-25.
[http://dx.doi.org/10.1002/med.21572] [PMID: 30887551]

[32] Gill J, Hyman P. Phage choice, isolation, and preparation for phage therapy. Curr Pharm Biotechnol 2010; 11(1): 2-14.
[http://dx.doi.org/10.2174/138920110790725311] [PMID: 20214604]

[33] Petrovic Fabijan A, Lin RCY, Ho J, *et al.* Safety of bacteriophage therapy in severe *Staphylococcus aureus* infection. Nat Microbiol 2020; 5(3): 465-72.
[http://dx.doi.org/10.1038/s41564-019-0634-z] [PMID: 32066959]

[34] Bruttin A, Brüssow H. Human volunteers receiving *Escherichia coli* phage T4 orally: a safety test of phage therapy. Antimicrob Agents Chemother 2005; 49(7): 2874-8.
[http://dx.doi.org/10.1128/AAC.49.7.2874-2878.2005] [PMID: 15980363]

[35] World Health Organization. 2020 Antibacterial agents in clinical and preclinical development: an overview and analysis. Geneva 2021.

[36] Huang DB, Sader HS, Rhomberg PR, Gaukel E, Borroto-Esoda K. Anti-staphylococcal lysin, LSVT-1701, activity: *In vitro* susceptibility of *Staphylococcus aureus* and coagulase-negative staphylococci (CoNS) clinical isolates from around the world collected from 2002 to 2019. Diagn Microbiol Infect Dis 2021; 101(3)115471
[http://dx.doi.org/10.1016/j.diagmicrobio.2021.115471] [PMID: 34280671]

[37] Abedon ST, Kuhl SJ, Blasdel BG, Kutter EM. Phage treatment of human infections. Bacteriophage 2011; 1(2): 66-85.
[http://dx.doi.org/10.4161/bact.1.2.15845] [PMID: 22334863]

[38] Ryan E, Garland MJ, Singh TRR, *et al.* Microneedle-mediated transdermal bacteriophage delivery. Eur J Pharm Sci 2012; 47(2): 297-304.
[http://dx.doi.org/10.1016/j.ejps.2012.06.012] [PMID: 22750416]

[39]   López-Jácome E, Franco-Cendejas R, Quezada H, *et al.* The race between drug introduction and appearance of microbial resistance. Current balance and alternative approaches. Curr Opin Pharmacol 2019; 48: 48-56.
[http://dx.doi.org/10.1016/j.coph.2019.04.016] [PMID: 31136908]

[40]   Christopher JL. Antimicrobial Resistance Collaborators. Global burden of bacterial antimicrobial resistance in 2019: a systematic analysis. Lancet 2022; 399(10325): 629-55.
[http://dx.doi.org/10.1016/S0140-6736(21)02724-0] [PMID: 35065702]

[41]   García-Quintanilla M, Pulido MR, López-Rojas R, Pachón J, McConnell MJ. Emerging therapies for multidrug resistant Acinetobacter baumannii. Trends Microbiol 2013; 21(3): 157-63.
[http://dx.doi.org/10.1016/j.tim.2012.12.002] [PMID: 23317680]

[42]   Bonilla N, Rojas MI, Netto Flores Cruz G, Hung SH, Rohwer F, Barr JJ. Phage on tap–a quick and efficient protocol for the preparation of bacteriophage laboratory stocks. PeerJ 2016; 4e2261.
[http://dx.doi.org/10.7717/peerj.2261] [PMID: 27547567]

[43]   Schooley RT, Biswas B, Gill JJ, *et al.* Development and use of personalized bacteriophage-based therapeutic cocktails to treat a patient with a disseminated resistant *Acinetobacter baumannii* infection. Antimicrob Agents Chemother 2017; 61(10)e00954-17
[http://dx.doi.org/10.1128/AAC.00954-17] [PMID: 28807909]

[44]   Van Belleghem JD, Dąbrowska K, Vaneechoutte M, Barr JJ, Bollyky PL. Interactions between bacteriophage, bacteria, and the mammalian immune system. Viruses 2018; 11(1): 10.
[http://dx.doi.org/10.3390/v11010010] [PMID: 30585199]

[45]   Burrowes BH, Molineux IJ, Fralick JA. Directed *in vitro* evolution of therapeutic bacteriophages: The appelmans protocol. Viruses 2019; 11(3): 241.
[http://dx.doi.org/10.3390/v11030241] [PMID: 30862096]

[46]   Van Nieuwenhuyse B, Van der Linden D, Chatzis O, *et al.* Bacteriophage-antibiotic combination therapy against extensively drug-resistant Pseudomonas aeruginosa infection to allow liver transplantation in a toddler. Nat Commun 2022; 13(1): 5725.
[http://dx.doi.org/10.1038/s41467-022-33294-w] [PMID: 36175406]

[47]   Eskenazi A, Lood C, Wubbolts J, *et al.* Combination of pre-adapted bacteriophage therapy and antibiotics for treatment of fracture-related infection due to pandrug-resistant *Klebsiella pneumoniae.* Nat Commun 2022; 13(1): 302.
[http://dx.doi.org/10.1038/s41467-021-27656-z] [PMID: 35042848]

[48]   Nir-Paz R, Gelman D, Khouri A, *et al.* Successful Treatment of Antibiotic-resistant, Poly-microbial Bone Infection With Bacteriophages and Antibiotics Combination. Clin Infect Dis 2019; 69(11): 2015-8.
[http://dx.doi.org/10.1093/cid/ciz222] [PMID: 30869755]

[49]   Aslam S, Courtwright AM, Koval C, *et al.* Early clinical experience of bacteriophage therapy in 3 lung transplant recipients. Am J Transplant 2019; 19(9): 2631-9.
[http://dx.doi.org/10.1111/ajt.15503] [PMID: 31207123]

[50]   Ferry T, Kolenda C, Laurent F, *et al.* Personalized bacteriophage therapy to treat pandrug-resistant spinal *Pseudomonas aeruginosa* infection. Nat Commun 2022; 13(1): 4239.
[http://dx.doi.org/10.1038/s41467-022-31837-9] [PMID: 35869081]

[51]   Fadlallah A, Chelala E, Legeais JM. Corneal infection therapy with topical bacteriophage administration. Open Ophthalmol J 2015; 9(1): 167-8.
[http://dx.doi.org/10.2174/1874364101509010167] [PMID: 26862360]

[52]   Wittebole X, De Roock S, Opal SM. A historical overview of bacteriophage therapy as an alternative to antibiotics for the treatment of bacterial pathogens. Virulence 2014; 5(1): 226-35.
[http://dx.doi.org/10.4161/viru.25991] [PMID: 23973944]

[53]   Lin DM, Koskella B, Lin HC. Phage therapy: An alternative to antibiotics in the age of multi-drug

resistance. World J Gastrointest Pharmacol Ther 2017; 8(3): 162-73.
[http://dx.doi.org/10.4292/wjgpt.v8.i3.162] [PMID: 28828194]

[54]   Eaton MD, Bayne-Jones S. Bacteriophage therapy. J Am Med Assoc 1934; 103(23): 1769.
[http://dx.doi.org/10.1001/jama.1934.72750490003007]

[55]   Górski A, Wazna E, Dąbrowska BW, Dabrowska K, Switała-Jeleń K, Miedzybrodzki R. Bacteriophage translocation. FEMS Immunol Med Microbiol 2006; 46(3): 313-9.
[http://dx.doi.org/10.1111/j.1574-695X.2006.00044.x] [PMID: 16553803]

[56]   Borysowski J, Górski A. Is phage therapy acceptable in the immunocompromised host? Int J Infect Dis 2008; 12(5): 466-71.
[http://dx.doi.org/10.1016/j.ijid.2008.01.006] [PMID: 18400541]

[57]   Tetz G, Tetz V. Bacteriophage infections of microbiota can lead to leaky gut in an experimental rodent model. Gut Pathog 2016; 8(1): 33.
[http://dx.doi.org/10.1186/s13099-016-0109-1] [PMID: 27340433]

[58]   Skurnik M, Pajunen M, Kiljunen S. Biotechnological challenges of phage therapy. Biotechnol Lett 2007; 29(7): 995-1003.
[http://dx.doi.org/10.1007/s10529-007-9346-1] [PMID: 17364214]

[59]   Philipson C, Voegtly L, Lueder M, et al. Characterizing phage genomes for therapeutic applications. Viruses 2018; 10(4): 188.
[http://dx.doi.org/10.3390/v10040188] [PMID: 29642590]

[60]   De Sordi L, Khanna V, Debarbieux L. The gut microbiota facilitates drifts in the genetic diversity and infectivity of bacterial viruses. Cell Host Microbe 2017; 22(6): 801-808.e3.
[http://dx.doi.org/10.1016/j.chom.2017.10.010] [PMID: 29174401]

[61]   Chhibber S, Kaur S, Kumari S. Therapeutic potential of bacteriophage in treating *Klebsiella pneumoniae* B5055-mediated lobar pneumonia in mice. J Med Microbiol 2008; 57(12): 1508-13.
[http://dx.doi.org/10.1099/jmm.0.2008/002873-0] [PMID: 19018021]

[62]   Suh GA, Lodise TP, Tamma PD, et al. Considerations for the use of phage therapy in clinical practice. Antimicrob Agents Chemother 2022; 66(3)e02071-21
[http://dx.doi.org/10.1128/aac.02071-21] [PMID: 35041506]

[63]   Remick DG. Pathophysiology of Sepsis. Am J Pathol 2007; 170(5): 1435-44.
[http://dx.doi.org/10.2353/ajpath.2007.060872] [PMID: 17456750]

[64]   Liu D, Van Belleghem JD, de Vries CR, et al. The safety and toxicity of phage therapy: A review of animal and clinical studies. Viruses 2021; 13(7): 1268.
[http://dx.doi.org/10.3390/v13071268] [PMID: 34209836]

[65]   Zhang L, Hou X, Sun L, et al. *Staphylococcus aureus* bacteriophage suppresses LPS-induced inflammation in MAC-T bovine mammary epithelial cells. Front Microbiol 2018; 9: 1614.
[http://dx.doi.org/10.3389/fmicb.2018.01614] [PMID: 30083140]

[66]   Park K, Cha KE, Myung H. Observation of inflammatory responses in mice orally fed with bacteriophage T7. J Appl Microbiol 2014; 117(3): 627-33.
[http://dx.doi.org/10.1111/jam.12565] [PMID: 24916438]

CHAPTER 3

The Revival of Natural Products in Developing New Antimicrobial Drugs: An Opportunity for Sustainable Management

Israel Castillo-Juárez[1,*] and **José Luis Díaz-Nuñez[2]**

[1] *Institute of Basic Sciences and Engineering, Secihti-Autonomous University of the State of Hidalgo, Hidalgo, México*

[2] *Postgraduate in Botany, Comecyt–College of Postgraduates, Mexico*

Abstract: The global health crisis caused by antimicrobial resistance presents a challenge that demands a rapid response. For decades, the development of antibiotics has significantly improved our quality of life. Shortly after their discovery, diseases and pests that had caused numerous deaths were controlled. Similarly, antibiotics facilitated advancements in medical and surgical practices, as well as enhanced food supply through agricultural applications. The marketed antibiotics were developed based on research into natural products of microbial origin. Therefore, this chapter examines various ongoing efforts and strategies to discover new antibacterials from natural products. It explores both traditional methods and innovative techniques for managing non-culturable microorganisms. The results inspire hope for a second golden age for these molecules. However, we must exercise greater responsibility in their use.

Keywords: Antimicrobials, Antibiotics, Natural products, Non-culturable organisms, RiPPs, iChip technology.

INTRODUCTION

Microorganisms are the first living beings to appear on Earth and have evolved for over three billion years. However, we became aware of their existence only after Anton van Leeuwenhoek discovered them 348 years ago [1]. Likewise, the eighteenth century was a period of essential advances in microbiology, of which the association of some microorganisms as etiological agents of diseases and early attempts to eliminate them are outstanding [2].

* **Corresponding author Israel Castillo-Juárez:** Institute of Basic Sciences and Engineering, Secihti-Autonomous University of the State of Hidalgo, Hidalgo, Mexico; E-mail: israel_castillo@uaeh.edu.mx

Mariano Martínez-Vázquez (Ed.)

One of the diseases that scientists have sought to eradicate is syphilis, an infectious disease caused by the *Treponema pallidum* bacterium that has coexisted with humans for thousands of years [3]. It is a disease that has caused contempt and fear due to its origins and devastating consequences. It has caused the death of millions of people and continues to affect many individuals today [4]. For centuries, without knowing the causative agent or understanding the biology of the disease, various treatments were developed in an attempt to cure it. One of the most popular remedies was the use of poisons, such as mercury salts, which acted nonspecifically, affecting both the microorganism and the patient [5].

It would not be until the beginning of the 20th century that Paul Ehrlich established the bases of antimicrobial chemotherapy with the revolutionary idea of "magic bullets," molecules aimed at eliminating disease-causing microorganisms without affecting the host cells [6]. This chapter aims to provide an overview of the discovery and nature of antibiotics, antibiotic resistance, and future perspectives, as well as new antimicrobials and new alternatives to reduce antimicrobial use.

AN ORGANOARSENIC COMPOUND

Salvarsan (compound 606) was Ehrlich's first magic bullet to treat syphilis; curiously, he developed it from organic compounds with the lead structure of arsenic, the favorite poison of medieval assassins [7]. Similarly, it should be noted that Salvarsan was initially developed to eliminate trypanosomes. Later, Ehrlich decided to evaluate its effect on the *T. pallidum* bacterium, as its discoverers, Schaudin and Hoffmann (1906), erroneously described it as a trypanosome-type parasite [8].

Although the effectiveness of salvarsan and the subsequent neosalvarsan (compound 914) in treating syphilis was controversial due to the side effects and deaths associated with its application, it is recognized that it kept the disease under control and saved the lives of millions [9].

It is essential to note that Salvarsan was used to understand the biology of the disease and its consequences in its late stages. In this regard, in 1913, Hideyo Noguchi managed to associate neurological disorders with *T. pallidum* infection by identifying them in the brains of patients with what would later be known as neurosyphilis [10].

THE RISE AND DECLINE OF ANTIBIOTICS

Despite the advances achieved with Salvarsan and Neosalvarsan, effective treatment against syphilis and many other diseases of microbial origin would not be possible until the discovery and development of penicillin by Alexander

Fleming in the 1940s [11]. These investigations, along with the development of synthetic sulfonamides in 1935 [12], gave rise to the discovery of antibiotics, one of humanity's most important scientific achievements [13]. In the 1950s, antibiotics ushered in a "golden age" in which the control and eradication of infectious diseases that had caused millions of deaths for centuries was achieved [14, 15].

Research on natural products made the discovery of antibiotics possible. By 1970, more than twenty classes were introduced to the market (Fig. **1**).

Fig. (1). Development of resistance to commercial antibiotics and some FDA-approved antimicrobials in recent years. Natural products of microbial origin were the basis for the discovery and development of commercial antibiotics in the "golden age" of the 1940s to the 1960s. Subsequent decades saw a sharp decline in the number of antibiotics entering the market. In addition, the pharmaceutical industry shifted its focus from discovering new classes of natural product antibiotics to synthesizing them from chemical libraries, which at the time were considered a more cost-effective strategy for developing new antibiotics. However, resistance is advancing rapidly, and chemical scaffolds from natural products are becoming obsolete, as microorganisms previously exposed to them have already developed strategies to generate resistance in a shorter time.

However, along with these discoveries came the warning that if the "magic bullets" were misused, they would lose their effectiveness due to the induction of bacterial resistance [16]. We have not taken this warning seriously for decades and favored the rapid spread of multidrug-resistant organisms [17]. Given this situation, various authors point out that we are entering the beginning of what has been defined as the post-antibiotic era, in which there are resistant microbial strains against which antibiotics are no longer effective [18]. Thus, an urgent call has been made to the scientific community to expedite the development of new antimicrobials [19].

The World Health Organization has declared antibiotic resistance a global public health problem, pointing to a group of six bacteria under the acronym "ESKAPE" (*Enterococcus faecium, Staphylococcus aureus, Klebsiella pneumoniae, Acinetobacter baumannii, Pseudomonas aeruginosa,* and *Enterobacter* species) for which the development of new antibacterial drugs is urgently required [20].

BULLETS WITH LITTLE MAGIC

On the other hand, the scientific advances made in recent decades related to studying microbiota and microbiomes have highlighted the increasing complexity of developing new antimicrobials [21]. Technological advancements in this field have shown that the diversity of the "microbial world" is broader and more complex than we knew, with only culturable microorganisms [22]. Currently, the so-called "microbial dark matter," microorganisms that cannot yet be cultured with current technology, plays a fundamental role in human health [23]. In addition, it may be a new source of bioactive metabolites to be explored [24].

Also, one of the limitations of not culturing pathogenic microorganisms is that it is impossible to carry out Koch's postulates unless the pathogenic agent is Isolated. A representative example is *Helicobacter pylori* and its relationship with various gastrointestinal diseases [25]. For decades, gastritis, peptic ulcer (gastric and duodenal), and gastric cancer were considered diseases of the modern era due to the drastic increase in their incidence in the mid-20th century [26].

They were even considered idiopathic diseases with a high rate of lifelong recurrence [27]. Various factors were deemed responsible, including stress, the consumption of irritants, or medications such as non-steroidal anti-inflammatory drugs [27]. Therefore, treatment was mainly focused on using antacids or protective substances for the gastric mucosa [27].

Since 1893, pathologists have observed the presence of curved bacilli in gastric biopsies of patients. They were proposed as the causal agents of these diseases, but they could not be cultured [28]. It would not be until 1980 that Australian researchers Robin Warren and Barry Marshall found the proper growing conditions for the bacterium, managed to isolate it, and demonstrated Koch's postulates after Marshall was inoculated with the culture and developed the disease [29]. For showing that these diseases have an infectious origin, they received the Nobel Prize in Medicine and Physiology in 2005 [30]. In addition, they revolutionized the treatment, which is currently based on schemes with at least two antibiotics to eliminate the bacteria from the stomach [31].

However, we are still far from a happy ending to this story. Although antibiotic treatment has reduced the prevalence of lower gastric diseases (peptic ulcer and gastric cancer), the indiscriminate eradication of *H. pylori* has been associated with an increase in upper gastric disorders, such as Barrett's syndrome and esophageal cancer, among others [32, 33]. Also, it has been shown that *H. pylori* has co-evolved with humans since its origins on the African continent and serves as an excellent marker of human migration [34]. However, how the bacterium acquired a pathogenic relationship with humans is still debatable. A possible explanation could be the massive use of antibiotics, mainly in the mid-20th century, which generated dysbiosis in the *H. pylori* populations in the stomach.

Similarly, alterations or dysbiosis caused in the human microbiota have been associated with several diseases, including obesity, cancer, depression, and autism [35]. In this sense, there is evidence that "classic antibiotics" generate dysbiosis by affecting the natural microbiota of individuals [36], so the design of new "magic bullets" now must also consider how not to affect the patient's microbiota.

NATURAL PRODUCTS ARE THE PRIMARY SOURCE OF ANTIBIOTICS

The production of new antibiotics has significantly decreased between 1960 and 2000; since 2000, no record of a new class has been introduced to the market (Fig. **1**) [37]. Likewise, although some new antibiotics have been approved by the Food and Drug Administration (FDA), such as linezolid, daptomycin, eravacycline, or combinations such as meropenem-vaborbactam [38], the appearance of resistance in a short time is reported (Fig. **1**). A similar situation occurs with some fluoroquinolones, such as moxifloxacin, approved in 1999, to which resistance was reported in 2021 [39 - 41]. A similar situation occurred with delafloxacin, approved in 2017; resistance was reported in 2020 [42, 43].

Various studies have documented the importance of natural products and their derivatives in discovering and developing new antimicrobials [44]. A recent study reported that, of 1,394 small molecules approved from 1981 to 2019, 67% are related to natural products (natural origin, derivative, or synthetic derivative), while only 33% are synthetically engineered molecules [45].

Natural products, particularly those of microbial origin, have been essential in discovering several bioactive compounds [46]. The culturable actinobacteria group stands out, contributing more than 90% of the antibiotics on the market [47]. Within actinobacteria, species of the genus *Streptomyces* are one of the primary sources of secondary metabolites, including bactericides and antiparasitics such as ivermectin and its derivatives [48] (Fig. **2**). The interest and usefulness of this genus are evident, as evidenced by the large number of culturable species characterized (more than 500), most of which belong to the Domain of bacteria [49].

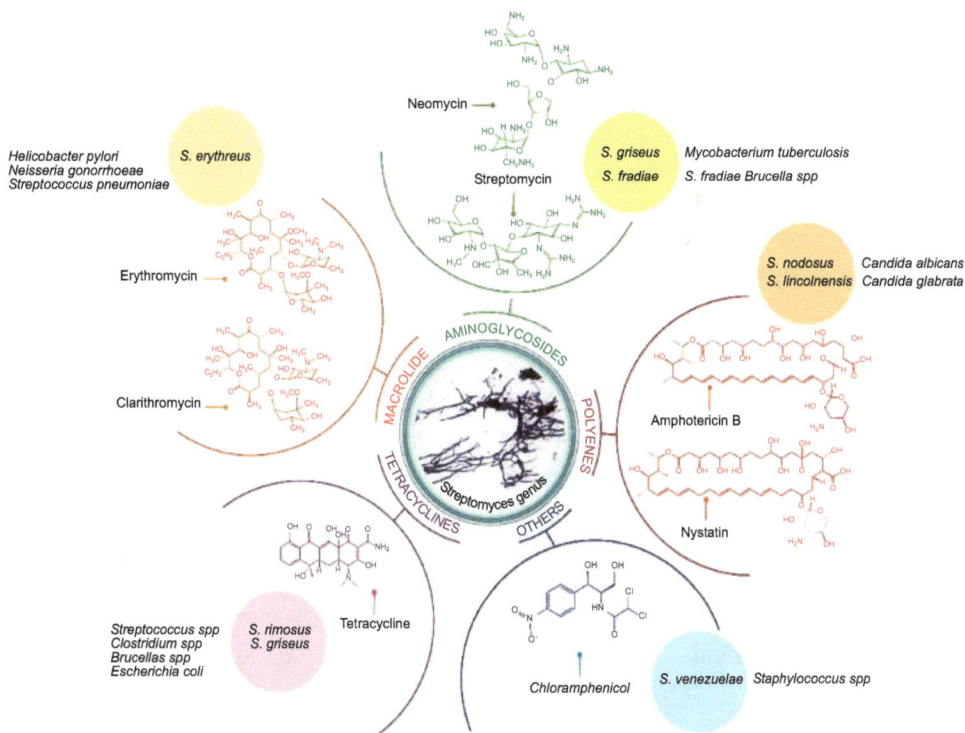

Fig. (2). The main classes of commercial antibiotics are isolated from culturable species of the Streptomyces genus. Many have been used as lead structures to make derivatives or to design synthetic molecules. Some commercial antibiotics have also been shown to be used against representative microbial pathogens. A representative micrograph of *S. erythraeus* is shown in the center of the image.

Natural products have served as chemical scaffolds for developing synthetic derivatives, with which antimicrobials have been enhanced and made safer [50]. However, before exposure to these antibiotics, the microorganisms already had mechanisms that favored the development of resistance to the new derivatives in less time [46]. Thus, resistance has been rapidly detected in new FDA-approved antimicrobials with great potential for short-term clinical use.

CURRENT DEVELOPMENTS IN ANTIBIOTIC DISCOVERY

Natural cephalosporins were isolated from the fungus *Cephalosporium acremonium* in 1948, while cephalosporin C was isolated from the genus *Acremonium* in the 1960s. However, several synthetic analogs have been developed that have enhanced their bactericidal activity and generated cephalosporins grouped into at least five generations [51].

Cefiderocol is a new cephalosporin (β-lactam) with siderophore properties and acts as an iron chelator (Fig. **3A**) [52, 53]. It inhibits Gram-negative cell wall synthesis by binding to penicillin-binding proteins. Also, by acting as a siderophore, it enters the periplasmic space, where it is more stable when subjected to the action of β-lactamases and the efflux pumps [54]. For this reason, it is suggested that cefiderecol would have a low rate of resistance generation since it evades the main resistance mechanisms known for β-lactams [55]. This characteristic makes it an antibiotic with a high potential to be introduced into clinical practice. In 2019, the FDA approved the treatment of urinary tract infections [56].

The main change in its structure compared to other cephalosporins, such as ceftazidime and cefepime, is that it contains a catechol group in the side chain (Fig. **3A**) [57]. It has good bactericidal activity against Gram-negatives, such as *Pseudomonas aeruginosa*, *Acinetobacter baumannii*, *Burkholderia cepacia*, *Klebsiella pneumoniae*, and *Stenotrophomonas maltophilia* [55, 58]. It has even better activity than combination treatments, such as ceftazidime-avibactam or ceftazidime-meropenem [59]. Also, its efficacy has been demonstrated in murine infection models with multidrug-resistant strains, and there is evidence of its effectiveness in clinical trials, with minor side effects or side effects similar to those of other cephalosporins [54, 59]. However, recent analyses have identified resistance mechanisms related to the simultaneous expression of multiple β-lactamases in combination with permeability changes [56] and case reports of resistance in treatments [60, 61].

On the other hand, the development of omics-based bioinformatics tools and strategies in the last decade has opened up immense possibilities for non-culturable organisms [62]. Genome mining makes it possible to detect silent biosynthetic genes, while metagenomics, synthetic biology, and iChip technology make it possible to study non-culturable microorganisms [63]. These new approaches have expanded the field of bioprospecting, in addition to avoiding replication (rediscovery of known compounds) and reducing the rapid generation of bacterial resistance caused by previous exposure to similar chemical structures [64].

Ribosomally synthesized and post-translationally modified peptides (RiPPs) are biologically relevant natural products, including antimicrobials [65]. Many RiPPs have been identified through genome mining, including new groups of Biosynthetic Gene Clusters (BGC) and their chemical products [66]. Many new natural products cannot be identified because they cannot be produced under laboratory conditions [62]. They may require specific signals within complex communities for their expression, so BGCs are often cryptic or silent [67].

Darobactin (Fig. **3B**) is a RiPP found in the genomes of symbiotic bacteria associated with entomopathogenic nematodes of the genus *Photorhabdus* [68]. This peptide affects bacterial cell membranes by interfering with the BamA protein and is active against Gram-negative pathogens, such as *K. pneumoniae, P. aeruginosa*, and *E. coli* [68, 69]. Until now, no investigations of resistance to aerobactin have been reported; however, trials of their effect *in vitro* and *in vivo* continue to be carried out to generate the best synthetic candidate for clinical use [69, 70].

Lugdunin (Fig. **3C**) is a thiazolidine-containing cyclic peptide antibiotic BGCs transcribed from the *Staphylococcus lugdunensis* genome. Lugdunin promotes competition against Gram-positive pathogens, such as *Staphylococcus aureus,* and stimulates the host's immune response [71]. However, *S. aureus* strains with some degree of resistance to lugdunin have been reported [72].

Lactocillin (Fig. **3D**) is a thiopeptide-type RiPP identified in the genome of *Lactobacillus gasseri* [73]. It affects bacterial ribosomes, inhibiting protein synthesis, and is efficient against Gram-positive vaginal pathogens [73]. There are no reports of bacterial resistance or continuity of its effect to date.

Another thiopeptide antibiotic is **cutamycin** (Fig. **3E**), which was discovered as a BGC in the *Propionibacterium acnes* genome. It exhibits good inhibitory activity against bacteria of the *Staphylococcus* genus; however, its mechanism of action is unknown, and so far, there are no reports of resistance [74].

Fidaxomicin (Fig. **3F**) is another tiacumicin-type RiPP identified in the *Dactylosporangium aurantiacum* genome, which acts by inhibiting bacterial RNA polymerase and is used to treat infections caused by *Clostridioides difficile* [75, 76]. In 2019, resistant *C. difficile* strains were identified [77]. Therefore, to mitigate resistance, it is currently combined with myxopyronin B, an antibiotic isolated from a myxobacterium called *Myxococcus fulvus* [76, 78].

Myxobacteria are a group of prokaryotes with multicellular behavior and complex life cycles, while their genomes encode many Polyketide Synthases (PKS) and Non-Ribosomal Peptide Synthetases (NRPS) [79, 80]. **Odilorhabdin** (Fig. **3G**) is a peptide-type antibiotic synthesized by PKS and NRPS genes of myxobacteria found in symbiosis with entomopathogenic nematodes of the genera *Xenorhabdus, Nematophila*, and *Photorhabdus* [70]. The mechanism of action of odilorhabdins involves interference with protein synthesis at the bacterial ribosome level of the Enterobacteriaceae Family and against some Gram-positive bacteria [81, 82]. A recent study reported that strains of *Klebsiella pneumoniae* resistant to colistin are also resistant to odilorhabdins [83].

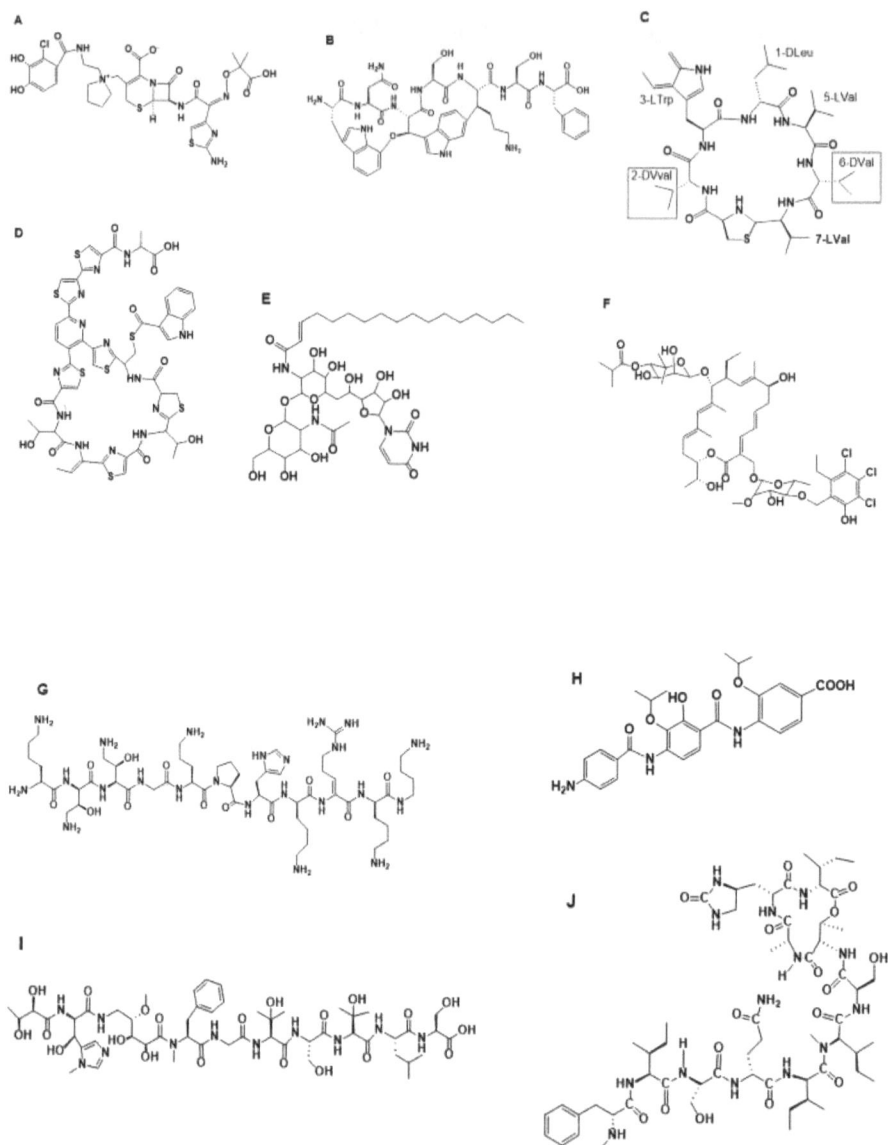

Fig. (3). Representative examples of new antibiotics related to studying natural antimicrobial products. A, cefiderocol; B, aerobactin; C, lugdunin; D, lactobacilli; E, cutamycin; F, fidaxomicin; G, odilorhabdin (NOSO-95C); H, cystobactamids (507); I, corramycin, and J, teixobactin.

Cystobactamids (Fig. **3H**) and **corramycin** (Fig. **3I**) are also produced by NRPS and PK of bacteria of the order Myxobacteriales [70]. Cystobactamids inhibit bacterial gyrase and type II topoisomerase and exhibit antimicrobial activity in pathogenic Gram-negative and Gram-positive bacteria [84, 85]. Although the corramycin mechanism of action is unknown, it is efficient against Gram-negative bacteria of the Enterobacteriaceae Family and against *A. baumannii* [86]. It should be noted that there are no reports of resistance to this group of antibiotics.

On the other hand, some outstanding examples of identifying new antibiotics in non-culturable representatives outside the laboratory have also been obtained using iChip technology [70, 87]. This novel tool consists of a small device with porous membranes that can separate a single class of bacteria from an entire community. It allows it to grow in its natural environment [87]. In this way, the **teixobactin** (Fig. **3J**) (an antibiotic of a new family) produced by *Eleftheria terrae* was identified. This antibiotic acts on multiple targets that block the synthesis of cell wall lipids [88]. It is effective against multi-resistant bacteria, such as *S. aureus*, *Clostridium difficile*, and *Mycobacterium tuberculosis*, among others. So far, there are no reports of resistance, and it has been suggested that the risk of generating resistance is low because its targets are highly specialized and conserved, such as undecaprenyl pyrophosphate, lipid I, and lipid II [70, 89].

TOWARDS THE SUSTAINABLE DISCOVERY AND DEVELOPMENT OF NEW ANTIBIOTICS

Bacterial resistance is a phenomenon inherent to the evolutionary process of microorganisms, which is why it most likely appeared from the beginnings of life on Earth [90]. However, the speed of appearance has been accelerated by human intervention, mainly due to the introduction of antibiotics to the clinic and their intensive use in agriculture [91]. A primary example of the misuse of "magic bullets" is that more than 70% of the current production of antibiotics has non-therapeutic uses, as they are used to induce weight gain in animals in the livestock industry [91]. Also, resistance genes are spread all over the planet [15], even in the polar caps, so curbing resistance is already a significant challenge [92].

Most new naturally occurring approved antibiotics have been discovered in terrestrial environments [93]; thus, culturable soil microorganisms are the primary source of antimicrobial compounds. However, with modern techniques being developed for managing non-culturable organisms, the panorama for identifying new bioactive compounds is expanding [63]. Other little-explored environments, such as the marine environment, which represent a promising source of new natural antimicrobial products, should be addressed [94].

No one doubts that we urgently need new antibiotics for treating infectious diseases. However, responsible use must also be promoted to avoid the mass extinction of other microbial species and the severe consequences that this generates for human health [95, 96].

For this reason, it is essential to investigate antimicrobial options other than the biocidal effect. One of these strategies is using antivirulence molecules, whose mechanism of action is to interfere with specific targets responsible for the establishment and pathogenicity of microorganisms [97]. It should be noted that a key property of this class of antimicrobials is that they do not directly interfere with bacterial viability, suggesting that they do not generate resistance or will do so at a slower rate [98]. One of the most outstanding antivirulence targets is quorum sensing, a cellular communication mechanism used by bacteria, archaea, and some unicellular eukaryotes to coordinate the expression of virulence factors [98, 99].

In this sense, it has been suggested that most of the molecules that we use as antibiotics have a different ecological-environmental role, as their concentration is naturally too low to achieve a biocidal effect [100]. Plants have had a more discreet role in identifying bactericidal molecules. Still, it is proposed that this may be because antibacterial defense mechanisms mainly involve antivirulence [14].

CONCLUSION

Natural products undoubtedly remain a promising source of antimicrobial compounds. Furthermore, there is a high probability of rapidly discovering new, more potent bactericidal compounds, but this will be helpful only if we refrain from misusing these potent molecules. Although we put forward the idea that we are in a period of resurgence of this kind of study, we have always looked for solutions in natural products. However, if a second "golden age" of antibiotics emerges, we will have a new opportunity to use them more responsibly now.

ACKNOWLEDGEMENTS

Israel Castillo-Juárez was supported by the I x M program, Secihti (CIR/030/2024). Díaz-Nuñez J.L. was supported by the Cátedra-Comecyt program (RCAT2024-0003).

REFERENCES

[1]   Leeuwenhoek AV. Observations, communicated to the publisher by Mr. Antony van Leeuwenhoek in a Dutch letter on the 9th of October. Here English'd: concerning little animals by him observed in rain, well, sea, and snow water; as also in water wherein pepper had lain infused Philos Trans R Soc Lond 12(133): 821-1.

[http://dx.doi.org/10.1098/rstl.1677.0003]

[2]   Grzybowski A, Pietrzak K. Robert Koch (1843-1910) and dermatology on his 171st birthday. Clin Dermatol 2014; 32(3): 448-50.
[http://dx.doi.org/10.1016/j.clindermatol.2013.10.005] [PMID: 24887990]

[3]   Malyarchuk AB, Andreeva TV, Kuznetsova IL, *et al.* Genomics of Ancient Pathogens: First Advances and Prospects. Biochemistry (Mosc) 2022; 87(3): 242-58.
[http://dx.doi.org/10.1134/S0006297922030051] [PMID: 35526849]

[4]   Denman J, Hodson J, Manavi K. Infection risk in sexual contacts of syphilis: A systematic review and meta-analysis. J Infect 2022; 84(6): 760-9.
[http://dx.doi.org/10.1016/j.jinf.2022.04.024] [PMID: 35447230]

[5]   Carocci A, Rovito N, Sinicropi MS, Genchi G. Mercury toxicity and neurodegenerative effects. Rev Environ Contam Toxicol 2014; 229: 1-18.
[PMID: 24515807]

[6]   Gelpi A, Gilbertson A, Tucker JD. Magic bullet: Paul Ehrlich, Salvarsan and the birth of venereology. Sex Transm Infect 2015; 91(1): 68-9.
[http://dx.doi.org/10.1136/sextrans-2014-051779]

[7]   Ferrie JE. Arsenic, antibiotics and interventions. Int J Epidemiol 2014; 43(4): 977-82.
[http://dx.doi.org/10.1093/ije/dyu152] [PMID: 25237690]

[8]   De Kruif PH, Novy RL. Anaphylatoxin and Anaphylaxis. I. Trypanosome Anaphylatoxin. J Infect Dis 1917; 20(5)

[9]   Swain K. 'Extraordinarily arduous and fraught with danger': syphilis, Salvarsan, and general paresis of the insane. Lancet Psychiatry 2018; 5(9): 702-3.
[http://dx.doi.org/10.1016/S2215-0366(18)30221-9] [PMID: 29866584]

[10]   Tan SY, Furubayashi J. Hideyo Noguchi (1876-1928): Distinguished bacteriologist. Singapore Med J 2014; 55(10): 550-1.
[http://dx.doi.org/10.11622/smedj.2014140] [PMID: 25631898]

[11]   Christensen SB. Drugs that changed society: History and current status of the early antibiotics: Salvarsan, sulfonamides, and β-lactams. Molecules 2021; 26(19): 6057.
[http://dx.doi.org/10.3390/molecules26196057] [PMID: 34641601]

[12]   Jacoby GA. History of drug-resistant microbes. Antimicrobial Drug Resistance: Mechanisms of Drug Resistance 1: 3-8.
[http://dx.doi.org/10.1007/978-3-319-46718-4_1]

[13]   Hutchings MI, Truman AW, Wilkinson B. Antibiotics: past, present and future. Curr Opin Microbiol 2019; 51: 72-80.
[http://dx.doi.org/10.1016/j.mib.2019.10.008] [PMID: 31733401]

[14]   Díaz-Nuñez JL, García-Contreras R, Castillo-Juárez I. The new antibacterial properties of the plants: Quo vadis studies of anti-virulence phytochemicals? Front Microbiol 2021; 12: 667126.
https://www.frontiersin.org/articles/10.3389/fmicb.2021.667126/full
[http://dx.doi.org/10.3389/fmicb.2021.667126] [PMID: 34025622]

[15]   Murray CJL, Ikuta KS, Sharara F, *et al.* Global burden of bacterial antimicrobial resistance in 2019: a systematic analysis. Lancet 2022; 399(10325): 629-55.
[http://dx.doi.org/10.1016/S0140-6736(21)02724-0] [PMID: 35065702]

[16]   Letek M. Alexander Fleming, the discoverer of the antibiotic effects of penicillin. Front Young Minds 2020; 7: 159.
[http://dx.doi.org/10.3389/frym.2019.00159]

[17]   Prescott JF. The resistance tsunami, antimicrobial stewardship, and the golden age of microbiology. Vet Microbiol 2014; 171(3-4): 273-8.

[http://dx.doi.org/10.1016/j.vetmic.2014.02.035] [PMID: 24646601]

[18]   Hansson K, Brenthel A. Imagining a post-antibiotic era: a cultural analysis of crisis and antibiotic resistance. Med Humanit 2022; 48(3): 381-8.
[http://dx.doi.org/10.1136/medhum-2022-012409] [PMID: 35922118]

[19]   Price R. O'Neill report on antimicrobial resistance: funding for antimicrobial specialists should be improved. Eur J Hosp Pharm Sci Pract 2016; 23(4): 245-7.
[http://dx.doi.org/10.1136/ejhpharm-2016-001013] [PMID: 31156859]

[20]   World   Health   Organization.   2021.   https://iris.who.int/bitstream/handle/10665/354545/9789240047655-eng.pdf?sequence=1

[21]   Dixit K, Chaudhari D, Dhotre D, Shouche Y, Saroj S. Restoration of dysbiotic human gut microbiome for homeostasis. Life Sci 2021; 278: 119622.
[http://dx.doi.org/10.1016/j.lfs.2021.119622] [PMID: 34015282]

[22]   Berdy B, Spoering AL, Ling LL, Epstein SS. *In situ* cultivation of previously uncultivable microorganisms using the ichip. Nat Protoc 2017; 12(10): 2232-42.
[http://dx.doi.org/10.1038/nprot.2017.074] [PMID: 29532802]

[23]   Vigneron A, Cruaud P, Guyoneaud R, Goñi-Urriza M. Into the darkness of the microbial dark matter *in situ* activities through expression profiles of *Patescibacteria* populations. Front Microbiol 2023; 13: 1073483.
[http://dx.doi.org/10.3389/fmicb.2022.1073483] [PMID: 36699594]

[24]   Sinha R, Sharma B, Dangi AK, Shukla P. Recent metabolomics and gene editing approaches for synthesis of microbial secondary metabolites for drug discovery and development. World J Microbiol Biotechnol 2019; 35(11): 166.
[http://dx.doi.org/10.1007/s11274-019-2746-2] [PMID: 31641867]

[25]   Salvatori S, Marafini I, Laudisi F, Monteleone G, Stolfi C. *Helicobacter pylori* and Gastric Cancer: Pathogenetic Mechanisms. Int J Mol Sci 2023; 24(3): 2895.
[http://dx.doi.org/10.3390/ijms24032895] [PMID: 36769214]

[26]   Graham DY. History of *Helicobacter pylori*, duodenal ulcer, gastric ulcer, and gastric cancer. 2014.https://www.ncbi.nlm.nih.gov/pmc/articles/PMC4017034/

[27]   Lam SK. Antacids: The past, the present, and the future. Baillieres Clin Gastroenterol 1988; 2(3): 641-54.
[http://dx.doi.org/10.1016/S0950-3528(88)80011-3] [PMID: 3048455]

[28]   Kidd M, Modlin IM. A century of *Helicobacter pylori*: paradigms lost-paradigms regained. Digestion 1998; 59(1): 1-15.
[http://dx.doi.org/10.1159/000007461] [PMID: 9468093]

[29]   Marshall BJ, Armstrong JA, McGechie DB, Clancy RJ. Attempt to fulfil Koch's postulates for pyloric Campylobacter. Med J Aust 1985; 142(8): 436-9.
[http://dx.doi.org/10.5694/j.1326-5377.1985.tb113443.x] [PMID: 3982345]

[30]   Van Der Weyden MB, Armstrong RM, Gregory AT. The 2005 Nobel Prize in physiology or medicine. Med   J   Aust   2005;   183(11-12):   612-4. https://www.mja.com.au/system/files/issues/183_11_051205/van11000_fm.pdf
[http://dx.doi.org/10.5694/j.1326-5377.2005.tb00052.x] [PMID: 16336147]

[31]   Alfarouk KO, Bashir AHH, Aljarbou AN, *et al.* The possible role of *Helicobacter pylori* in gastric cancer and its management. Front Oncol 2019; 9: 75.
[http://dx.doi.org/10.3389/fonc.2019.00075] [PMID: 30854333]

[32]   Reshetnyak VI, Burmistrov AI, Maev IV. *Helicobacter pylori* : Commensal, symbiont or pathogen? World J Gastroenterol 2021; 27(7): 545-60. https://www.ncbi.nlm.nih.gov/pmc/articles/PMC7901052/
[http://dx.doi.org/10.3748/wjg.v27.i7.545] [PMID: 33642828]

[33] Blaser MJ. In a world of black and white, *Helicobacter pylori* is gray. Ann Intern Med 1999; 130(8): 695-7.
[http://dx.doi.org/10.7326/0003-4819-130-8-199904200-00019] [PMID: 10215569]

[34] Linz B, Balloux F, Moodley Y, *et al.* An African origin for the intimate association between humans and *Helicobacter pylori*. Nature 2007; 445(7130): 915-8. https://www.nature.com/articles/nature05562
[http://dx.doi.org/10.1038/nature05562] [PMID: 17287725]

[35] Duvallet C, Gibbons S, Gurry T, Irizarry R, Alm E. Meta analysis of microbiome studies identifies shared and disease-specific patterns. bioRxiv 2017; 134031. https://www.biorxiv.org/content/10.1101/134031v1 abstract

[36] Ramirez J, Guarner F, Bustos Fernandez L, Maruy A, Sdepanian VL, Cohen H. Antibiotics as major disruptors of gut microbiota. Front Cell Infect Microbiol 2020; 10: 572912.
[http://dx.doi.org/10.3389/fcimb.2020.572912] [PMID: 33330122]

[37] Coates A, Hu Y, Bax R, Page C. The future challenges facing the development of new antimicrobial drugs. Nat Rev Drug Discov 2002; 1(11): 895-910.https://www.nature.com/articles/nrd940
[http://dx.doi.org/10.1038/nrd940] [PMID: 12415249]

[38] Yusuf E, Bax HI, Verkaik NJ, van Westreenen M. An update on eight "new" antibiotics against multidrug-resistant gram-negative bacteria. J Clin Med 2021; 10(5): 1068.
[http://dx.doi.org/10.3390/jcm10051068] [PMID: 33806604]

[39] König E, Ziegler HP, Tribus J, Grisold AJ, Feierl G, Leitner E. Surveillance of antimicrobial susceptibility of anaerobe clinical isolates in southeast austria: Bacteroides fragilis group is on the fast track to resistance. Antibiotics (Basel) 2021; 10(5): 479.
[http://dx.doi.org/10.3390/antibiotics10050479] [PMID: 33919239]

[40] Dewangan V, Sahu RK, Satapathy T. Incidence of Moxifloxacin serious adverse drug reactions in Pneumococcal infections virus infected patients detected by a Pharmacovigilance program by laboratory signals in a Tertiary hospital in Chhattisgarh (India) Res J Pharmacology Pharmacodynamics 2022; 14(4): 237-45. https://www.indianjournals.com/ijor.aspx?target=ijor:rjppd&volume=14&issue=4&article=007

[41] Paukner S, Moran GJ, Sandrock C, *et al.* A plain language summary of how lefamulin alone can be used to treat pneumonia caught outside of the hospital due to common bacterial causes, including drug-resistant bacteria. Future Microbiol 2022; 17(6): 397-410.
[http://dx.doi.org/10.2217/fmb-2021-0276] [PMID: 35285291]

[42] Iregui A, Khan Z, Malik S, Landman D, Quale J. Emergence of delafloxacin-resistant Staphylococcus aureus in Brooklyn, New York. Clin Infect Dis 2020; 70(8): 1758-60.
[http://dx.doi.org/10.1093/cid/ciz787] [PMID: 31412357]

[43] Kocsis B, Gulyás D, Szabó D. Delafloxacin, finafloxacin, and zabofloxacin: novel fluoroquinolones in the antibiotic pipeline. Antibiotics (Basel) 2021; 10(12): 1506.
[http://dx.doi.org/10.3390/antibiotics10121506] [PMID: 34943718]

[44] Katz L, Baltz RH. Natural product discovery: past, present, and future. J Ind Microbiol Biotechnol 2016; 43(2-3): 155-76.
[http://dx.doi.org/10.1007/s10295-015-1723-5] [PMID: 26739136]

[45] Newman DJ, Cragg GM. Natural products as sources of new drugs over the nearly four decades from 01/1981 to 09/2019. J Nat Prod 2020; 83(3): 770-803. https://pubs.acs.org/doi/full/10.1021/acs.jnatprod.9b01285
[http://dx.doi.org/10.1021/acs.jnatprod.9b01285] [PMID: 32162523]

[46] Schneider YK. Bacterial natural product drug discovery for new antibiotics: strategies for tackling the problem of antibiotic resistance by efficient bioprospecting. Antibiotics (Basel) 2021; 10(7): 842.
[http://dx.doi.org/10.3390/antibiotics10070842] [PMID: 34356763]

[47] Jose PA, Jha B. New dimensions of research on actinomycetes: quest for next generation antibiotics.

Front Microbiol 2016; 7: 1295.
[http://dx.doi.org/10.3389/fmicb.2016.01295] [PMID: 27594853]

[48]   Bérdy J. Thoughts and facts about antibiotics: Where we are now and where we are heading. J Antibiot (Tokyo) 2012; 65(8): 385-95. https://www.nature.com/articles/ja201227
[http://dx.doi.org/10.1038/ja.2012.27] [PMID: 22511224]

[49]   Ruminococcus ET. Bergey's Manual of Systematics of Archaea and Bacteria 2015; 1-5.
[http://dx.doi.org/10.1016/j.drudis.2015.01.009]

[50]   Patridge E, Gareiss P, Kinch MS, Hoyer D. An analysis of FDA-approved drugs: natural products and their derivatives. Drug Discov Today 2016; 21(2): 204-7.
[http://dx.doi.org/10.1016/j.drudis.2015.01.009] [PMID: 25617672]

[51]   Bui T, Patel P, Preuss CV. Cephalosporins. In: StatPearls. Treasure Island (FL): StatPearls Publishing 2025. https://www.ncbi.nlm.nih.gov/books/NBK551517/

[52]   Sato T, Yamawaki K. Cefiderocol: discovery, chemistry, and in vivo profiles of a novel siderophore cephalosporin. Clin Infect Dis 2019; 69 (Suppl. 7): S538-43.
[http://dx.doi.org/10.1093/cid/ciz826] [PMID: 31724047]

[53]   El-Lababidi RM, Rizk JG. Cefiderocol: a siderophore cephalosporin. Ann Pharmacother 2020; 54(12): 1215-31.
[http://dx.doi.org/10.1177/1060028020929988] [PMID: 32522005]

[54]   Jorda A, Zeitlinger M. Pharmacological and clinical profile of cefiderocol, a siderophore cephalosporin against gram-negative pathogens. Expert Rev Clin Pharmacol 2021; 14(7): 777-91.
[http://dx.doi.org/10.1080/17512433.2021.1917375] [PMID: 33849355]

[55]   Canton R, Doi Y, Simner PJ. Treatment of carbapenem-resistant *Pseudomonas aeruginosa* infections: a case for cefiderocol. Expert Rev Anti Infect Ther 2022; 20(8): 1077-94.
[http://dx.doi.org/10.1080/14787210.2022.2071701] [PMID: 35502603]

[56]   Karakonstantis S, Rousaki M, Kritsotakis EI. Cefiderocol: systematic review of mechanisms of resistance, heteroresistance and *in vivo* emergence of resistance. Antibiotics (Basel) 2022; 11(6): 723.
[http://dx.doi.org/10.3390/antibiotics11060723] [PMID: 35740130]

[57]   McCarthy MW. Cefiderocol to treat complicated urinary tract infection. Drugs Today (Barcelona) 2020; 56(3): 177–84.
[http://dx.doi.org/10.1358/dot.2020.56.3.3118466]

[58]   Aoki T, Yoshizawa H, Yamawaki K, *et al.* Cefiderocol (S-649266), A new siderophore cephalosporin exhibiting potent activities against Pseudomonas aeruginosa and other gram-negative pathogens including multi-drug resistant bacteria: Structure activity relationship. Eur J Med Chem 2018; 155: 847-68.
[http://dx.doi.org/10.1016/j.ejmech.2018.06.014] [PMID: 29960205]

[59]   Zhanel GG, Golden AR, Zelenitsky S, *et al.* Cefiderocol: a siderophore cephalosporin with activity against carbapenem-resistant and multidrug-resistant gram-negative bacilli. Drugs 2019; 79(3): 271-89.
[http://dx.doi.org/10.1007/s40265-019-1055-2] [PMID: 30712199]

[60]   Takemura M, Yamano Y, Matsunaga Y. Characterization of shifts in minimum inhibitory concentrations during treatment with cefiderocol or comparators in the Phase 3 CREDIBLE-CR and APEKS-NP studies Open Forum Infect Dis 2020; 7(Suppl_1): S649-50.
[http://dx.doi.org/10.1093/ofid/ofaa439.1449]

[61]   Meschiari M, Volpi S, Faltoni M, *et al.* Real-life experience with compassionate use of cefiderocol for difficult-to-treat resistant *Pseudomonas aeruginosa* (DTR-P) infections. JAC Antimicrob Resist 2021; 3(4): dlab188.
[http://dx.doi.org/10.1093/jacamr/dlab188] [PMID: 34909691]

[62]   Kloosterman AM, Medema MH, van Wezel GP. Omics-based strategies to discover novel classes of

RiPP natural products. Curr Opin Biotechnol 2021; 69: 60-7.
[http://dx.doi.org/10.1016/j.copbio.2020.12.008] [PMID: 33383297]

[63]   Lodhi AF, Zhang Y, Adil M, Deng Y. Antibiotic discovery: combining isolation chip (iChip) technology and co-culture technique. Appl Microbiol Biotechnol 2018; 102(17): 7333-41.
[http://dx.doi.org/10.1007/s00253-018-9193-0] [PMID: 29974183]

[64]   Maghembe R, Damian D, Makaranga A, *et al.* Omics for bioprospecting and drug discovery from bacteria and microalgae. Antibiotics (Basel) 2020; 9(5): 229.
[http://dx.doi.org/10.3390/antibiotics9050229] [PMID: 32375367]

[65]   Li Y, Rebuffat S. The manifold roles of microbial ribosomal peptide–based natural products in physiology and ecology. J Biol Chem 2020; 295(1): 34-54.
[http://dx.doi.org/10.1074/jbc.REV119.006545] [PMID: 31784450]

[66]   Blin K, Kim HU, Medema MH, Weber T. Recent development of antiSMASH and other computational approaches to mine secondary metabolite biosynthetic gene clusters. Brief Bioinform 2019; 20(4): 1103-13.
[http://dx.doi.org/10.1093/bib/bbx146] [PMID: 29112695]

[67]   Rutledge PJ, Challis GL. Discovery of microbial natural products by activation of silent biosynthetic gene clusters. Nat Rev Microbiol 2015; 13(8): 509-23. https://www.nature.com/articles/nrmicro3496
[http://dx.doi.org/10.1038/nrmicro3496] [PMID: 26119570]

[68]   Imai Y, Meyer KJ, Iinishi A, *et al.* A new antibiotic selectively kills Gram-negative pathogens. Nature 2019; 576(7787): 459-64. https://www.nature.com/articles/s41586-019-1791-1
[http://dx.doi.org/10.1038/s41586-019-1791-1] [PMID: 31747680]

[69]   Kaur H, Jakob RP, Marzinek JK, *et al.* The antibiotic darobactin mimics a β-strand to inhibit outer membrane insertase. Nature 2021; 593(7857): 125-9. https://www.nature.com/articles/s41586-02--03455-w
[http://dx.doi.org/10.1038/s41586-021-03455-w] [PMID: 33854236]

[70]   Walesch S, Birkelbach J, Jézéquel G, *et al.* Fighting antibiotic resistance—strategies and (pre)clinical developments to find new antibacterials. EMBO Rep 2023; 24(1): e56033.
[http://dx.doi.org/10.15252/embr.202256033] [PMID: 36533629]

[71]   Zipperer A, Konnerth MC, Laux C, *et al.* Human commensals producing a novel antibiotic impair pathogen colonization. Nature 2016; 535(7613): 511-6. https://www.nature.com/articles/nature18634
[http://dx.doi.org/10.1038/nature18634] [PMID: 27466123]

[72]   Krauss S, Zipperer A, Wirtz S, *et al. et-al.* Secretion of and self-resistance to the novel fibupeptide antimicrobial lugdunin by distinct ABC transporters in *Staphylococcus lugdunensis*. Antimicrob Agents Chemother 2020; 65(1): e01734-20.
[http://dx.doi.org/10.1128/AAC.01734-20] [PMID: 33106269]

[73]   Donia MS, Cimermancic P, Schulze CJ, *et al.* A systematic analysis of biosynthetic gene clusters in the human microbiome reveals a common family of antibiotics. Cell 2014; 158(6): 1402-14.
[http://dx.doi.org/10.1016/j.cell.2014.08.032] [PMID: 25215495]

[74]   Claesen J, Spagnolo JB, Ramos SF, *et al.* A *Cutibacterium acnes* antibiotic modulates human skin microbiota composition in hair follicles. Sci Transl Med 2020; 12(570): eaay5445.
[http://dx.doi.org/10.1126/scitranslmed.aay5445] [PMID: 33208503]

[75]   Aldape MJ, Packham AE, Heeney DD, Rice SN, Bryant AE, Stevens DL. Fidaxomicin reduces early toxin A and B production and sporulation in *Clostridium difficile in vitro*. J Med Microbiol 2017; 66(10): 1393-9.
[http://dx.doi.org/10.1099/jmm.0.000580] [PMID: 28893366]

[76]   Brauer M, Herrmann J, Zühlke D, Müller R, Riedel K, Sievers S. Myxopyronin B inhibits growth of a Fidaxomicin-resistant *Clostridioides difficile* isolate and interferes with toxin synthesis. Gut Pathog 2022; 14(1): 4.

[http://dx.doi.org/10.1186/s13099-021-00475-9] [PMID: 34991700]

[77]   Schwanbeck J, Riedel T, Laukien F. Characterization of a clinical *Clostridioides difficile* isolate with markedly reduced fidaxomicin susceptibility and a V1143D mutation in rpoB Journal of Antimicrobial Hemotherapy 74(1): 6-10.
[http://dx.doi.org/10.1093/jac/dky375]

[78]   Irschik H, Gerth K, Höfle G, Kohl W, Reichenbach H. The myxopyronins, new inhibitors of bacterial RNA synthesis from *Myxococcus fulvus* (Myxobacterales). J Antibiot (Tokyo) 1983; 36(12): 1651-8.
[http://dx.doi.org/10.7164/antibiotics.36.1651] [PMID: 6420386]

[79]   Schäberle TF, Goralski E, Neu E, *et al.* Marine myxobacteria as a source of antibiotics--comparison of physiology, polyketide-type genes and antibiotic production of three new isolates of *Enhygromyxa salina*. Mar Drugs 2010; 8(9): 2466-79.
[http://dx.doi.org/10.3390/md8092466] [PMID: 20948900]

[80]   Schäberle TF, Lohr F, Schmitz A, König GM. Antibiotics from myxobacteria. Nat Prod Rep 2014; 31(7): 953-72.
[http://dx.doi.org/10.1039/c4np00011k] [PMID: 24841474]

[81]   Pantel L, Florin T, Dobosz-Bartoszek M, *et al.* Odilorhabdins, antibacterial agents that cause miscoding by binding at a new ribosomal site. Mol Cell 2018; 70(1): 83-94.e7.
[http://dx.doi.org/10.1016/j.molcel.2018.03.001] [PMID: 29625040]

[82]   Racine E, Gualtieri M. From worms to drug candidate: the story of odilorhabdins, a new class of antimicrobial agents. Front Microbiol 2019; 10: 2893.
[http://dx.doi.org/10.3389/fmicb.2019.02893] [PMID: 31921069]

[83]   Pantel L, Juarez P, Serri M, *et al.* Missense mutations in the CrrB protein mediate odilorhabdin derivative resistance in *Klebsiella pneumoniae*. Antimicrob Agents Chemother 2021; 65(5): e00139-21.
[http://dx.doi.org/10.1128/AAC.00139-21] [PMID: 33685902]

[84]   Herrmann J, Lukežič T, Kling A. *et al.* Strategies for the discovery and development of new antibiotics from natural products: three case studies Curr Top Microbiol Immunol 2016; 398: 339-63.
[http://dx.doi.org/10.1007/82_2016_498]

[85]   Groß S, Schnell B, Haack PA, Auerbach D, Müller R. *In vivo* and *in vitro* reconstitution of unique key steps in cystobactamid antibiotic biosynthesis. Nat Commun 2021; 12(1): 1696.
[http://dx.doi.org/10.1038/s41467-021-21848-3] [PMID: 33727542]

[86]   Couturier C, Groß S, von Tesmar A. *et-al.* Structure elucidation, biosynthesis, total synthesis and antibacterial in-vivo efficacy of myxobacterial Corramycin. ChemRxiv 2022.
[http://dx.doi.org/10.26434/chemrxiv-2022-97gp2]

[87]   Ling LL, Schneider T, Peoples AJ, *et al.* A new antibiotic kills pathogens without detectable resistance. Nature 2015; 517(7535): 455-9.
[http://dx.doi.org/10.1038/nature14098] [PMID: 25561178]

[88]   McCarthy MW. Teixobactin: a novel anti-infective agent. Expert Rev Anti Infect Ther 2019; 17(1): 1-3.
[http://dx.doi.org/10.1080/14787210.2019.1550357] [PMID: 30449226]

[89]   Karas JA, Chen F, Schneider-Futschik EK, *et al.* Synthesis and structure–activity relationships of teixobactin. Ann N Y Acad Sci 2020; 1459(1): 86-105.
[http://dx.doi.org/10.1111/nyas.14282] [PMID: 31792983]

[90]   D'Costa VM, King CE, Kalan L, *et al.* Antibiotic resistance is ancient. Nature 2011; 477(7365): 457-61.
[http://dx.doi.org/10.1038/nature10388] [PMID: 21881561]

[91]   Serratosa J, Blass A, Rigau B, *et al.* Residues from veterinary medicinal products, growth promoters and performance enhancers in food-producing animals: a European Union perspective. Rev Sci Tech

2006; 25(2): 637-53.
[http://dx.doi.org/10.20506/rst.25.2.1687] [PMID: 17094703]

[92]   Fitzpatrick D, Walsh F. Antibiotic resistance genes across a wide variety of metagenomes. FEMS Microbiol Ecol 2016; 92(2): fiv168.
[http://dx.doi.org/10.1093/femsec/fiv168] [PMID: 26738556]

[93]   Hegemann JD, Birkelbach J, Walesch S, Müller R. Current developments in antibiotic discovery. EMBO Rep 2023; 24(1): e56184.
[http://dx.doi.org/10.15252/embr.202256184] [PMID: 36541849]

[94]   Choudhary A, Naughton L, Montánchez I, Dobson A, Rai D. Current status and future prospects of marine natural products (MNPs) as antimicrobials. Mar Drugs 2017; 15(9): 272.
[http://dx.doi.org/10.3390/md15090272] [PMID: 28846659]

[95]   Blaser MJ, Falkow S. What are the consequences of the disappearing human microbiota? Nat Rev Microbiol 2009; 7(12): 887-94.
[http://dx.doi.org/10.1038/nrmicro2245] [PMID: 19898491]

[96]   Blaser MJ. The theory of disappearing microbiota and the epidemics of chronic diseases. Nat Rev Immunol 2017; 17(8): 461-3.
[http://dx.doi.org/10.1038/nri.2017.77] [PMID: 28749457]

[97]   García-Contreras R, Martínez-Vázquez M, González-Pedrajo B, Castillo-Juárez I. Editorial: Alternatives to Combat Bacterial Infections. Front Microbiol 2022; 13: 909866.
[http://dx.doi.org/10.3389/fmicb.2022.909866] [PMID: 35602022]

[98]   Castillo-Juárez I, Maeda T, Mandujano-Tinoco EA, *et al.* Role of quorum sensing in bacterial infections. World J Clin Cases 2015; 3(7): 575-98.

[99]   Sprague GF Jr, Winans SC. Eukaryotes learn how to count: quorum sensing by yeast: Figure 1. Genes Dev 2006; 20(9): 1045-9.
[http://dx.doi.org/10.1101/gad.1432906] [PMID: 16651650]

[100]   Aminov RI. The role of antibiotics and antibiotic resistance in nature. Environ Microbiol 2009; 11(12): 2970-88.
[http://dx.doi.org/10.1111/j.1462-2920.2009.01972.x] [PMID: 19601960]

CHAPTER 4

Carbapenemases: Impact, Perspectives, and Identification Methods

Luis Esau López Jacome[1,2,*] and **Rafael Franco Cendejas**[3]

[1] *Clinical Microbiology Laboratory, Infectious Diseases Division, National Institute of Rehabilitation Luis Guillermo Ibarra Ibarra, México City, México*

[2] *Biology Department, Chemistry Faculty, National Autonomous University of México, México City, México*

[3] *Subdirection of Biomedical Research, National Institute of Rehabilitation Luis Guillermo Ibarra Ibarra, México City, México*

Abstract: Multidrug resistance is a global and severe public health concern; according to data from Jim O´Neil, in 2050, deaths could rise to 10 million. Each time, clinical options are decreasing. The outlook is discouraging, not considering the increased resistance during the COVID-19 pandemic.

The last line of defense for clinicians is the use of carbapenems, and developing new clinical pharmacological strategies involves the use of site-specific inhibitors. Identifying the enzymes that degrade these types of antimicrobials is crucial. It is the key to reducing the selection pressure and ensuring their proper use. The half-life of a molecule is dictated by its prescription, and this is where the clinical microbiology laboratory comes in. It is not just a place for testing; it serves as the guiding axis for the correct, adequate, and rational administration of drugs. In this chapter, we will delve into the importance of carbapenemases-mediated resistance and its classification, impact, and detection strategies.

Keywords: Carbapenemase, Carbapenemase classification, Detection strategies, Global concern, Identification tools.

INTRODUCTION

Antimicrobials are molecules that increase life expectancy, being different before and after 1900. According to "Our World in Data," before 1900, life expectancy

* **Corresponding author Luis Esau López Jacome:** Clinical Microbiology Laboratory, Infectious Diseases Division, National Institute of Rehabilitation Luis Guillermo Ibarra Ibarra, México City, México and Biology Department, Chemistry Faculty, National Autonomous University of México, México City, México;
E-mail: esaulopezjacome@gmail.com

Mariano Martínez-Vázquez (Ed.)

was around 30-40 years; after 1900, life expectancy increased to the current expectancy [1]. For example, in 2015, life expectancy in the United States of America increased to 79 years, in Canada to 82 years, in Europe to around 80 years, and in Mexico to 77 years. The lowest age rates are identified in Africa (around 60 years) (Fig. **1**). These changes may be associated with introducing the first and subsequent antimicrobial molecules. The first molecule introduced to cure infectious diseases was a synthetic one, discovered by Paul Ehrlich's research group: Salvarsan, also known as the magic bullet, designed to treat syphilis. The following molecule, discovered by serendipity, was penicillin by Alexander Fleming in 1928 [2]. However, thanks to the work of H.W. Florey ed. al, penicillin reached mass production [3].

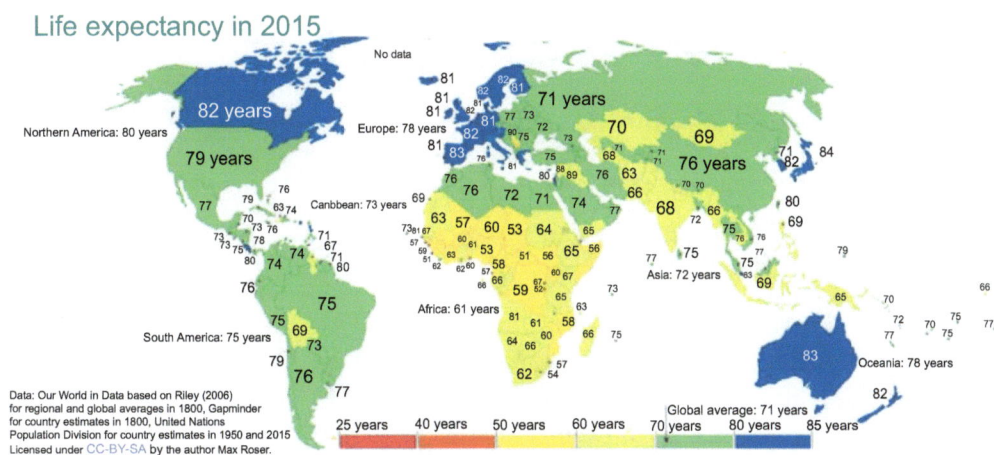

Fig. (1). Life expectancy in 2015 taken from https://ourworldindata.org/life-expectancy [1].

Despite new molecules being discovered as soon as they were announced, the response from microorganisms also began to appear. For example, resistance to penicillin appeared just a year after its commercial launch [4]. In response, the study of antimicrobial resistance began to gain strength as physicians had fewer medicines to treat patients' infections, and mortality rates associated with antimicrobial resistance increased. At this point, various governments and researchers worldwide showed interest in this topic. One of the most cited works is those belonging to Sir. Jim O'Neill. He has published various essays about the importance of addressing multidrug resistance. He has made projections about the impact of multidrug resistance until 2050. According to his work, this year, there will be more deaths associated with multidrug resistance than those with pathologies with mortality associated per se, such as cancer—it has been calculated that there will be approximately 10 million deaths [5]. The region with the highest death rate will be Asia, with 4,730,000 deaths attributable to antimicrobial drug resistance, followed by Africa, with 4,150,000 deaths. According to projections for the American continent, the number of deaths is expected to rise to 709,000. This data shows that the cumulative economic impact since 2014 will have a negative influence on the gross domestic product [6]. However, O'Neill's prediction seems inaccurate since, in January 2022, Dr. Mohsen Naghavi and collaborators published "The Global Burden of Bacterial Antimicrobial Resistance in 2019." In this paper, the authors estimated that 4.95 million deaths are associated with bacterial antimicrobial resistance. This estimate shows that in 2019, the deaths related to bacterial resistance were nearly half of the 10 million predicted for 2050. The highest rate of deaths attributable to bacterial antimicrobial resistance was observed in sub-Saharan Africa, with 27.3 deaths per 100,000 inhabitants, while Australia had the lowest rate [7].

In 2017, the World Health Organization (WHO) created a list of primary pathogens that require immediate attention to combat antimicrobial resistance. The microorganisms were prioritized based on their level of threat. Priority was given to infections associated with healthcare, such as *Enterobacterales*. These are resistant to third and fourth cephalosporin generations as well as carbapenems. *Acinetobacter baumannii and Pseudomonas aeruginosa*, which are resistant to carbapenems, were also prioritized [8].

In addition to the misuse of antibiotics, the COVID-19 pandemic is also a leading cause of pressure selection in microorganisms. On 11 March 2020, the WHO declared a pandemic that began in Wuhan and spread to the rest of the world [9]. Unfortunately, this disease was initially misdiagnosed as bacterial pneumonia and treated with antibiotic administration. As a result, antibiotic misuse and abuse contributed to the rapid rise of antimicrobial resistance. Several works have listed the pandemic's role in this phenomenon.

For example, in Mexico, the use of antibiotics was compared between the pre-pandemic and pandemic stages, resulting in increased resistance to oxacillin, erythromycin, and clindamycin for *Staphylococcus aureus,* as well as to imipenem and meropenem for *Klebsiella pneumoniae, as* determined in blood samples [10]. Before the pandemic, the Pan-American Health Organization reported that around 10-15% of antibiotics were administered for secondary infection. During the pandemic, antibiotics increased by around 94-100%, of which 72% were broad-spectrum [11].

On the other hand, in the USA, the Centers for Disease Control and Prevention (CDC) reported an alarming increase in resistant infections by around 15% from 2019 to 2020. According to their report, there was an increase of 78% in carbapenem-resistant *A. baumannii*, 32% more *P. aeruginosa* multidrug-resistant, and up to 35% in carbapenem-resistant *Enterobacterales*. The pandemic played a role in the increase of antimicrobial resistance [12].

CARBAPENEMS

Carbapenems are one of the last weapons physicians have for the treatment of infections caused primarily by Gram-negative, but also by Gram-positive bacteria [13].

Carbapenems are antibiotics with the characteristic of having a ß-lactam ring in the core of the molecule, and unlike penicillins, Carbapenem has no ring of sulfur. In contrast, the fused ring is unsaturated in C2, a position where a substituent is present [13, 14]. Some changes in the molecular configuration allow for better stability against ß-lactamases. For example, the methyl group in meropenem at the 1ß position increases stability against renal dehydropeptidase. In contrast, in imipenem, protection of molecular integrity must be achieved through cilastatin, an inhibitor of renal dehydropeptidase [15]. The first compound discovered carrying the structural backbone of Carbapenem was olivanic acid. However, thienamycin was the first Carbapenem and was the model for the next generation of carbapenems (Fig. **2**).

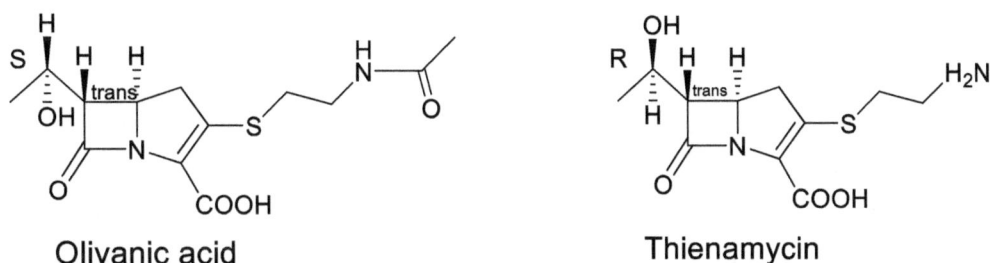

Fig. (2). Olivanic acid and Thienamycin structures.

The first molecules developed had specific characteristics. Olivanic acid was unstable and did not penetrate bacterial cells, while thienamycin was unstable in water and reacted strongly to nucleophiles [14, 16]. These molecules were complex to purify and produce on a large scale. Despite these limitations, thienamycin showed promise as an antibiotic, with activity against Gram-negative rods like *Pseudomonas aeruginosa*, anaerobes like *Bacteroides fragilis*, and bacteria like *Staphylococcus aureus* and streptococci [14, 17].

Thienamycin effectively treats various infections, including those caused by anaerobic bacteria and Extended-Spectrum ß-Lactamases (ESBL) producers [18, 19]. However, imipenem was more effective than thienamycin as it was less susceptible to degradation and had an affinity for PBPs, as well as stability against certain ß-lactamases. Imipenem, being a carbapenem, required the support of another molecule against the activity of renal dehydropeptidase, and thus, it was accompanied by cilastatin [14, 20]. Panipenem, similar to imipenem, also requires cilastatin to inhibit the action of renal enzymes, allowing imipenem to increase the concentration in urine [20] (Fig. **3**).

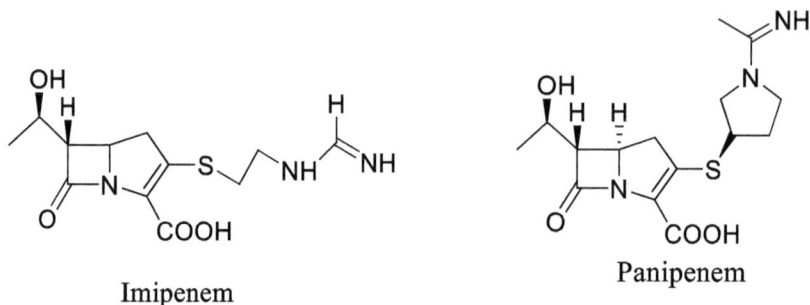

Fig. (3). Chemical structure of imipenem and panipenem.

New molecules were joining. However, the first concern was still to find a trick to evade enzymatic degradation mediated by renal dehydropeptidase; this was achieved through chemical derivatization by adding a methyl group into the 1ß position. This change was found to play a protective role against enzymatic activity, and as soon as this change was described, molecules with more stability were discovered or invented, such as ertapenem, doripenem, faropenem, and biapenem [14, 21, 22] (Fig. **4**).

The action mechanism of carbapenems is cell wall inhibition. Carbapenems cannot diffuse through the membrane quickly; regarding this point, these molecules require the help of some structural proteins embedded in the membrane, known as porins, which are frequently referred to as OMPs (outer membrane proteins). Once they have passed and arrived at the periplasmic space, they are acylated by Penicillin-Binding Proteins (PBPs) [14, 21 - 23]. These proteins play a vital role in the formation of peptidoglycan. Peptidoglycan is a polymer that is the barrier of cells and gives the structure and form to the bacterial cell. Thus, PBPs catalyze the polymerization of the glycan strands, referred to as trans glycosylation, and the cross-linking between chains (known as transpeptidation), and some PBPs are capable of hydrolyzing pentapeptides, attacking the last D-alanine (DD-carboxypeptidase) or hydrolyzing the peptide bond that connects to glycan strands (endopeptidase) [23, 24]. Bacteria have several PBPs; as soon as new genomes are sequenced, information is obtained. PBPs can be categorized into 1) high molecular mass (HMM) and 2) low molecular mass (LMM).

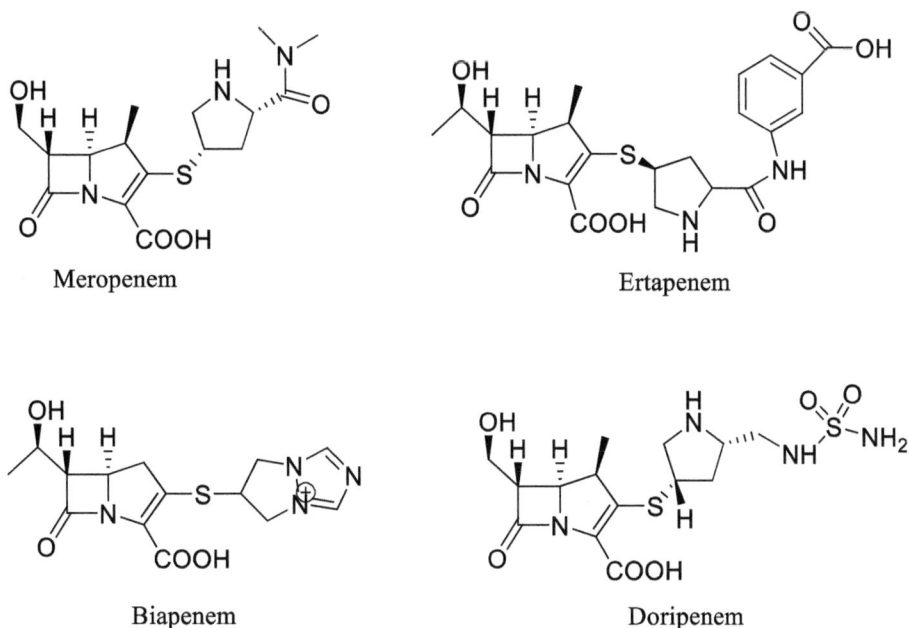

Fig. (4). Chemical structures of meropenem, ertapenem, biapenem, and doripenem.

On the other hand, depending on the catalytic activity and structure, it can also be divided into two classes. Class A is characterized by the glycosyltransferase activity associated with the N-terminal domain; this catalyzes the elongation of uncross-linked chains of glycans. Meanwhile, class B is thought to be associated with cell morphogenesis through its interaction with other proteins in the cell cycle [23, 25 - 27]. Besides these two classes, there is a third class, C. It is frequently described as LMM and subdivided into three groups, C1–C3; nonetheless, Eric Sauvage et al. s., in their work, described class C as four subgroups: type 4, type 5, type 7, and type AmpH. Traditionally, PBP numbering is according to SDS-PAGE migration, but in some cases, it is unclear since many of them (PBPs) have similar running patterns. For example, PBP2 of *S. aureus*, which belongs to class A, is similar to *Escherichia coli* PBP1a. Another is the case of *S. aureus* PBP1, which is identical to PBP3 of *E. coli* [23]. PBPs have several roles in peptidoglycan synthesis, primarily involved in the elongation and cell division processes. For example, in *E. coli*, 12 PBPs have been widely described and studied; from these, seven are class C, which is associated with recycling, cell separation, and peptidoglycan maturation.

Furthermore, it possesses class A: PBP1a, PBP1b, and PBP1c. PBP2 and PBP3 belong to class B. At the same time, PBP1a and PBP1b are defined as significant transpeptidases and transglycosylases. Information about PBP1c is limited, and its role is poorly understood [23, 28 - 31].

On the other hand, the two class B PBPs of *E. coli* are monofunctional transpeptidases. PBP2 is involved in elongation, defined as elongate due to its dynamic proteinic complex related explicitly to cell elongation, and PBP3 is a major protein of the divisome (the cell division complex) [23, 32].

Class A PBP is necessary for cell growth in most bacteria but not in all. The absence of PBP1a or PBP1b in *E. coli* is not lethal; however, the loss of PBP1 or PBP1b in the presence of PBP1c results in lethality for this microorganism. Another example is *Neisseria gonorrhoeae*, which has only one class A, PBP1; the result is reduced viability [24, 30, 33]. For class B, five subclasses have been described, from B1 to B5. The two prominent examples of B1 are PBP2a in *S. aureus* and PBP5 in *Enterococcus faecalis*; members in this subclass are supposed to have a reduced affinity for penicillin [24]. Elongase complex specific to PBP2 is contained in subclass B2 for Gram-negative bacteria, while the divisome is classified into subclass B3, particular to the PBP3 type of Gram-negative bacteria. The example of B4 is for PBP2x, found in Streptococcus pneumoniae and other enzymes involved in cell division (in Gram-positive bacteria), and subclass 5 is present in Gram-positive bacteria. Still, they are not involved directly in septation [24, 34 - 36]. Besides these PBP belonging to class B, others have been defined and found in mycobacteria, *Streptomyces*, and bacteria related to members of these subclasses referred to as Subclasses B-like I, II, and III [24, 27]. For class C, a type AmpH structure is based on the DD-peptidase of *Streptomyces*. It has three motifs, two of which are well-conserved. First, SXXK is like other PBPs. Second, conserved Serine (SXN) and the third, not well conserved in this family [KTG(T/S)], the complete enzymatic function is not yet well studied; however, as a curiosity, class C ß-lactamases are closely related to R61 (*Streptomyces*) [24].

Peptidoglycan is a molecule that weighs around 3×10^9 Da. It comprises 2.7×10^6 to 3.5×10^6 N-acetylglucosamine (Nacgluc)-N-acetylmuramic acid (Nacmur)-pentapeptide monomers per cell [37 - 39]. Peptidoglycan provides shape and stiffness to bacterial cells and helps them resist osmotic pressures [39]. The molecule consists of linear glycan strands crosslinked by short peptide chains. The strands contain Nacgluc and Nacmur linked by ß-1,4 glycosidic bonds [39 - 42]. In several Gram-negative bacteria, the pentapeptide chain is made up of L-alanine (L-ala), γ-D-glutamate (D-glu), mesodiaminopimelic acid (mDAP), and D-alanine (D-ala) with the bond between D-glu and mDAP (such an isopeptide linkage) [39, 43]. This arrangement allows for crosslinking with other peptide chains. In contrast, in Gram-positive bacteria, the crosslinking conforms to the basis of the characteristic mesh structure of peptidoglycan. These peptides are crosslinked directly with bridging peptides of varying amino acid lengths and composition [39].

The synthesis of peptidoglycan is a complex process that involves several steps. It begins with the formation of Uridine Diphosphate (UDP) with N-Acetylglucosamine (GlcNAc) and the synthesis of the Outer Membrane (OM). UDP-*GlcNAc* is derived from the intermediate fructose-6-phosphate, amido group donor L-glutamine, and UTP by the sequential action of enzymes Glms, GlmM, and GlmU. UDP-Nacmur is formed from UDP-*GlcNAc* and phosphoenolpyruvate by the activity of MurA and MurB [39, 44 - 51]. Since MurA is involved in the initial synthesis phases, it can be blocked by mimetic molecules like fosfomycin. The synthesis of peptides is nonribosomal and is added by ATP-dependent aminoacyl ligases. The first amino acid to be added is L-ala, which goes onto UDP-Nacmur thanks to the activity of MurC to form UDP-Nacmur-L-ala. Then, L-glu is added to UDP-Nacmur-L-ala by MurD. MurI participates in the racemization of D-glu to L-glu. Then, mDAP is added by MurE to form UDP-Nacmur tripeptide, and finally, MurF adds the D-ala-D-ala dimer. Some analogs, such as D-cycloserine, can inhibit the precursors that block ligases (DdlA or DdlB) [52 - 61].

Tripeptides can also be ligated to the peptidoglycan recycling directly onto UDP-NacMur by Mpl [39, 62]. MraY participates in forming lipid-I, transferring the phospho NacMur-pentapeptide moiety onto an IM (inner membrane) lipid carrier, undecaprenyl pyrophosphate (UndPP/C55 isoprenyl pyrophosphate). UppS synthesizes UndPP and, as a result of dephosphorylation allowed by any of IM-bound phosphatases (BacA, PgpB, YbjG, or LpxT) [39, 63 - 66], and this dephosphorylation could be mediated by the action of bacitracin, an antibiotic capable of inhibiting the synthesis of peptidoglycan [67]. Once lipid I has been formed, the next step is the synthesis of lipid II, which is mediated by adding one molecule of UDP-*GlcNAc* onto lipid I. The action of MurG controls this process; with these steps, peptidoglycan precursors in the inner membrane are complete. Another protein responsible for the transport of Und-PP-O-antigen [68, 69] in outer membrane biosynthesis is the MurJ, where lipid-II binding in the central cavity of this protein allows for alternating between cytoplasm-facing and periplasm-facing [39, 70, 71]. Meanwhile, MurJ has predicted, thanks to bioinformatics, another protein has been associated with the transport of lipid-II; in this case, transportation activity was observed *in vivo*, FtsW, which has been involved in cell division and showed activity of glycosyl transferase, this is required for glycan strand polymerization during the synthesis of septum [39, 72 - 75]. Then, lipid-II polymerization is mediated by two septal or sidewall synthesis reactions: 1) trans glycosylation of disaccharide to form glycan strands and 2) transpeptidation to form cross-bridges [39]. The next step is mediated by elongase, a protein group. For several years, the paradigm regarding the synthesis of peptidoglycan was centered on the fact that it was regulated by class A PBP, which belongs to the high molecular weight and has both Trans Glycosylase (TGase) and Transpeptidase (TPase) activities. However, recent publications show that RodA has significant TGase activity in elongase. The glycan polymerization mediated by RodA is consecutively followed by cross-linking of peptides by the TPase PBP2, a member of class B. The domain of D, D-transpeptidase cleaves the D-ala4-D-ala5 (action site of vancomycin, for example) bond in a donor peptide and, again, cross-links the D-ala4, the mDAP3 of an acceptor, allowing the formation of a 4-3 cross-link [39, 76 - 81]. In this point, glycopeptides, such as vancomycin, can inhibit the biosynthesis of peptidoglycan binding to D-ala4-D-ala5 and inhibiting the polymerization by steric hindrance [39, 82], meanwhile ß-lactams antibiotics such as penicillin, cephalosporins, and carbapenems are designed to inhibit at level of transpeptidase domain of all PBP by covalent interactions with the active site [23, 39].

The interactions between transmembrane domains of RodA and PBP2 allow activation of TGase activity of RodA by PBP2, and together have formed a complete peptidoglycan synthetic unit, considering all enzymatic requirements for peptidoglycan polymerization [39, 74, 79, 83 - 85]. The well-controlled interactions between each one of the members involved in peptidoglycan synthesis allow for a uniform distribution of newly synthesized peptidoglycan, dictated by the sites where it must be produced. A homolog of actin, MreB, is a cytoskeletal protein conserved in bacteria and participates in cell shape and mechanical integrity. Besides, this can polymerize into filaments under the cytoplasmic membrane, and its polymerization is ATP-dependent [39, 86 - 91].

MreB is responsible for giving the rod shape and maintaining its form due to the presence of short and discontinuous filaments along the central membrane curvature [39]. Besides, MreC binds and ties filaments of MreB directly to the membrane and simultaneously recruits MreD, the third member of the cytoskeleton [39, 86 - 90, 92, 93]. Then, the interaction between elements of the cytoskeleton and RodA-PBP2 is mediated by IM RodZ, the latest member of the Rod complex [39, 94 - 98]. The PBPab (redundant presence of PBPa or PBPb) are clamped in IM and periplasmic proteins through a cytoplasmic tail, an N-terminal TGase domain that polymerizes lipid-II into glycan strands, and a C-terminal D, D TPase that cross-links peptides [39, 99 - 101]. These two are believed to be redundant. However, each one has a particular localization and function. For example, PBP1a is mainly located along the sidewall and collaborates with PBP2, both of which play a leading role in elongation.

On the other hand, PBP1b is localized along the sidewall, diffusely, beside the septum; with this, its role in cell division is well-defined due to its interaction with divisional elements, such as FtsN and FtsW-PBP3 [39, 102 - 106]. Then, once we reviewed the processes involved in peptidoglycan and understood the biological importance of this element, researchers developed several strategies to 1) study the physiological processes in the bacterial cell and 2) design strategies against those bacteria able to develop a disease in human beings. One of these strategies, as mentioned above, is the ß-lactam group, which includes penicillins, cephalosporins, monobactams, and carbapenems; these last ones were also previously described. The presence of ß-lactam rings confers a high affinity to PBP, so the block synthesis of peptidoglycan is allowed. Nevertheless, microorganisms are fascinating entities because they can develop various strategies to evade aggressive behavior, thereby generating resistance mechanisms. These mechanisms are divided into two ways: 1) intrinsic, which means microorganisms have, in their chromosomes, the faculty of evading the action of the antibiotic, such as the expression of efflux pump, production of enzymes that degrade antibiotics (ß-lactamases), structural differences between Gram positives and Gram-negative (outer membrane, which is a hydrophobic barrier) [107] and 2) acquired resistance. This is a major global concern, as antibiotics are ineffective against some infections caused by resistant microorganisms. In this group, we can find enzymes acquired through Mobile Genetic Elements (MGE) (Table **1**).

Table 1. Some examples of resistance mechanisms acquired horizontally in Gram-negative rods.

MGE	Antibiotic affected	Name of the mechanism involved	Example	References
Plasmid	Cephalosporin 1-3 generation (ß-lactam)	AmpC	CMY	[108]
Plasmid	Cephalosporins (ß-lactam), monobactam	Extended Spectrum ß-Lactamases (ESBL)	TEM	[109]
Plasmid	Carbapenem	Carbapenemases	NDM	[110]
Plasmid	Aminoglycosides	16S rRNA methyltransferase	ArmA in *Klebsiella pneumoniae* or RmtA in *Pseudomonas aeruginosa*	[111]
Plasmid	MLS$_B$ (Macrolides, Lincosamides and Streptogramins B)	Ribosomal methylation	Erm	[112]
Plasmid	Sulfamethoxazole	Dihydropteroate synthetase	Sul	[113]
Plasmid	Several	Efflux pump	TMexCD1-TOprJ1	[114]

(Table 1) cont.....

MGE	Antibiotic affected	Name of the mechanism involved	Example	References
Transposons	Carbapenems	Carbapenemase	Tn21 (OXA)	[115]
Integrons	Sulfamethoxazole	Dihydropteroate synthetase	Sul	[116]

As mentioned previously [8], one of these mechanisms was classified as a priority by the WHO in 2017. For microorganisms with resistance to carbapenem, it is necessary to discriminate between enzymatic and non-enzymatic mechanisms. Enzymes able to degrade carbapenems are defined as carbapenemases.

The first ß-lactamase defined as a metalloprotein was the ß-lactamase II found in *Bacillus cereus* [117], followed by two strains described in *Bacteroides fragilis* [118] and one found in *Aeromonas hydrophila* expressed by the gene *cphA* [119]. Until 1993, when the first carbapenemase with clinical implications was described, the number, distribution, and type of these enzymes also increased worldwide. The first carbapenamase was NmcA, localized in the chromosome of *Enterobacter cloacae* [120]. This protein reduced imipenem's Minimal Inhibitory Concentrations (MIC) [120]. Little by little, other enzymes that could hydrolyze carbapenems began to appear. The next one, belonging to the KPC family, was named as such because it was identified for the first time in *Klebsiella pneumoniae* [121]; the second one to be described in this family was KPC2, found in *Klebsiella oxytoca* [122]. As soon as they were described, they began to appear as new ones, and each time, more and more spread until the point we are now. To better understand these enzymes, ß-lactamases, several authors proposed different ways to classify them. However, K. Bush, G. A. Jacoby, and A. proposed the most widely distributed and used classifications. A Medeiros is also referred to as the Bush-Jacoby-Medeiros classification. In this classification, ß-lactamases are grouped into four groups. It is a biochemical classification, also named functional classification, where each one of the elements interacts with the inhibitor, enzyme, and substrate, and its interaction behavior allows the classification. Group 1 belongs to cephalosporinases, which are not well inhibited by clavulanic acid. Group 2 is composed of penicillinases, cephalosporinases, and broad-spectrum ß-lactamases. Active site-directed ß-lactamases inhibitors inhibit members of this group, those with serine residues. Group 3, the main characteristic is that members of this group have a metallic ion as a cofactor in the hydrolytic core, which helps with hydrolytic activity. This cofactor is zinc, the reason why these enzymes are named metallo-β-lactamases and can hydrolyze penicillins, cephalosporins, and carbapenems [123]. However, all those that cannot be characterized are grouped in group 4.

Ambler's classification was described in 1980. Because this classification is based on the level of amino acid sequences, it is known as structural classification and is integrated by four classes from A to D. Class A, C, and D are known as serine ß-lactamase since they have in their hydrolytic pocket the presence of serine; these residues are responsible for the hydrolysis of ß-lactams.

Class A has the most enzymes, both ß-lactamases and carbapenemases. The most frequent, such as TEM, SHV, and CTX-M [124 - 126], belong to class A. According to ß-lactamase DataBase–Structure and function (http://www.bldb.eu/Enzymes.php) Class A is integrated by followed enzymes: ACI, AER, AFA, R39, ARL, AST, ASU1, AXC, BBI, BcI, BcIII, BCL, BEL, BES, BIC, BKC, BlaC, BlaP, BlaS, BOR, BPS, BRO, CAD, CAE, CARB, CblA, CBP, CdiA, CepA, CfxA, CGA, CIA, CKA, CKO, CM1, CME, CRP, CRH, CSP, CST, CTX-M, CumA, CzoA, CPA, DBA, DES, ERP, FAR, FEC, FLC, FONA, FPH, FRI, FTU, GES, GIL, GMA, GPA, GPC, GRI, HBL, HER, HMS, HugA, IMI, KBL, KLUA, KLUB, KLUC, KLUG, KLUS, KLUY, KPC, L2, LAP, LEN, LUS, LUT, MAB, MAL, MFO, MIA, MIN, MM3, MOM, NmcA, OHIO, OIH, OKP, ORF6, ORN, OXY, PAD, PAL, PAU, PbbA, PC1, PC2, PEC, PenA, PenB, PenBcc, PenE, PenF, PenG, PenH, PenI, PenJ, PenL, PenN, PenO, PER, PLA, PLES, PME, PSV, RAA, RAHN, RASA, RATA, RCP, REUT, RIC, ROB, RSA1, RSA2, RUB, SDA, SCO, SED, SFC, SFO, SGM, SHV, SME, SMO, TEM, TER, TLA, TLA2, VAN, VBR, VEB, VDA, VRB, XCC, and BlaA [127]. However, not all of them had carbapenemase activity. This class is composed of 1896 enzymes (Table **2**) [127].

Members of this class can hydrolyze aminopenicillins, ureidopenicillins, first and second-generation cephalosporins, and monobactams, and some of them have activity against carbapenems [127]. They are inhibited by clavulanate, tazobactam, and boronic acid derivatives [127, 128].

Class B has a heavyweight, the metallic enzymes, since they need the metallic cofactor to exhibit hydrolytic activity, which is why they are known as Metallo ß-lactamases. The massive advantage of metallo-β-lactamases is that they can practically hydrolyze all ß-lactam (except monobactam) and carbapenems [124, 126]. The members of this class have a completely different evolutionary origin from the serine ß-lactamases. The hydrolytic process of ß-lactams is also entirely different from the process of serine ß-lactamases [124, 125]. They maintain a zone highly conserved across the members of the superfamily of metallo-ß-lactamases; the common scaffold is an αß/ßα and metal binding motif. Members belonging to this class are unable to be inhibited by boronic acid derivatives and avibactam [129], and very few possess carbapenemase activity (3%) [130]. This class is divided into three subgroups (B1-B3), and the relevant enzymes from the clinical microbiology point of view are within B1. Meanwhile, B2 needs one ion of Zinc (Zn) for its hydrolysis, and B1 and B3 need two Zn ions [131]; however, it does not matter if it is B1, B2, or B3, all metallo-ß-lactamases share H/N116-X11- -H118-X119-D120-H/X121 in the N-terminal half of protein [130]. Subclasses are composed of B1 (594), B2 (24), and B3 (231) (Table **2**) [127].

Table 2. Integrative table with the two most common classifications.

Ambler	Bush-Jacoby-Medeiros	Characteristics	Inhibitor	Enzymes Representatives	Microorganisms
A	2a	Penicillinases	CA, TZB, SLB	PCI	*S. aureus*
	2b	Broad-spectrum enzymes	CA, TZB, SLB	TEM1, TEM2, SHV1	*Enterobacterales*
	2be	Extended spectrum ß-lactamases	CA, TZB, SLB	TEM3-28, SHV2-6, Toho-1, CTX-M, PER-1, VEB-1	*Enterobacterales, P. aeruginosa. A. baumannii*
	2br	Broad-spectrum resistant to inhibitors		TEM30-36, TRC-1	*E. coli, N. brasiliensis*
	2c	Carbenicillinases	CA, TZB, SLB	PSE-1, PSE-3, PSE-4, CARB-3, CARB-4	*P. aeruginosa*
	2e	Cephalosporinases	CA, TZB, SLB	FPM-1, CepA	*P. vulgaris, Bacteroides fragilis*
	2f	Carbapenemases	BA, DA	KPC1-3, IMI1, NMC-A, Sme1-2, GES-2	*K. pneumoniae, K. oxytoca, E. cloacae, S. marcescens*
B	3	Metallo ß-lactamase	EDTA, DA	VIM, IMP, NMD	*Aeromonas spp., Bacteroides fragilis, S. maltophila, Bacillus cereus, A. baumannii, K. pneumoniae*
C	1	Cephalosporinases	BA, Cloxacillin	AmpC, MIR-1, FOX-1, ACT-1, CMY-2	*E. cloacae, S. marcescens, C. freundii, A. baumannii, P. aeruginosa, E. coli*

(Table 2) cont.....

Ambler	Bush-Jacoby-Medeiros	Characteristics	Inhibitor	Enzymes Representatives	Microorganisms
D	2d	Cloxacillinase	Temocillin	OXA1-11, PSE-2 OXA-48, OXA-181	*Enterobacterales, P. aeruginosa, Acinetobacter spp.*
Unclassified	4	Penicillinase		Chr	*B. cepacia*

CA: clavulanic acid; TZB: tazobactam; SLB: sulbactam; EDTA: ethylenediaminetetraacetic acid; BA: boronic acid; DA: dipicolinic acid. Table adapted from previous studies [123, 132].

Subclass B1.

AFM, ANA, BcII, BIM, BlaB, CAM, CfiA, CGB, CEMC19, CrxA, CX1, DIM, EBR, ECV, ElBla2, FIA, FIM, GIM, GMB, GRD23, HBA, HMB, IMP, IND, JOHN, KHM, MOC, MUS, MYO, MYX, NDM, ORR, PAN, PEDO, PKB, PST, SFB, SHD, SHN, SIM, SLB, SPM, SPN79, SPS, STA, SZM, TTU, TMB, TUS, VAM, VIM, VMB, VMH, WUS, ZHO, and ZOG.

Subclass B2.

CphA, CVI, PFM, SFH, and YEM

Subclass B3.

AIM, ALG6, ALG11, AM1, BJP, BLEG, CAR, CAU, CHI, CPS, CRD3, CSR, DHT2, EAM, ECM, EFM, ELM, ESP, EVM, FEZ, GOB, L1, LMB, LRA2, LRA3, LRA7, LRA8, LRA12, LRA17, LRA19, MEMA1, MIM, MSI, NWM, PAM, PEDO, PJM, PLN, POM, PNGM, RM3, SAM, SER, SIE, SIQ, SMB, SPG, SPR, SSE, B3SU1, B3SU2, and THIN.

Class D.

Class D and class A are both serine ß-lactamases, with 1250 enzymes in this class. However, most of the enzymes with carbapenemase activity belong to the OXA group and are named OXA because they can break down isoxazolyl penicillins (such as oxacillin, methicillin, and cloxacillin) [131].

One of the distinguishing features of the OXA group is the high heterogeneity between its members, making it one of the most prominent families of enzymes. This heterogeneity means that no single molecule can identify all the members of the OXA family phenotypically. These enzymes can be found on both chromosomes and extra chromosomally.

In addition to some OXA, an enzyme with carbapenemase activity is also found in RAD [127, 133]. Unfortunately, this family is typically known for having a poor response to classical inhibitors, such as clavulanic acid, sulbactam, or tazobactam [131].

AFD, ATD, BAD, BAT, BED, BEN, BOC, BPU, BSD, BSU, CDD, CEMC18, CPD, LCR, NOD, NPS, OXA, RAD, RSD1, RSD2, and STD integrate class D.

IMPACT OF CARBAPENEMASE

Carbapenemase is a resistance mechanism that impacts patient outcomes. In 2021, the Centers for Disease Control and Prevention (CDC) tested 56,016 strains of *P. aeruginosa* resistant to carbapenems, out of which 2.11% had at least one codifying gene of carbapenemase (1,181 strains), with NDM being the primary enzyme involved [134]. The first enzyme that affected humans was discovered in 2001 by Yigit et al. This discovery marked a milestone in the subject of resistance. They described and characterized the first member of the KPC family found in a clinical isolate of *Klebsiella pneumoniae*, which is why it was named KPC (*Klebsiella pneumoniae* carbapenemase) [121]. The same group described the second in 2003 in a *Klebsiella oxytoca* [122]. The ß-lactamase Database - Structure and function has annotated 167 enzymes [127], and the reference gene catalog from the National Institutes of Health has annotated 161 [135]. In 2004, OXA-48 was described *in K. pneumoniae* in Turkey, and the third place was taken by NDM, a metallo-β-lactamase named after its discovery in New Delhi, and the M due is a metallo-β-lactamase [136].

Over time, more carbapenemases have emerged, raising concerns among the scientific, clinical, and microbiological communities due to the limited strategies to deal with them in infectious processes. Several articles have been written to draw attention to the impact of this resistance mechanism on patient outcomes, and we now know its effect. There is an association between mortality and carbapenemase, which can be seen in the work of Ruyin Zhou, a systematic review of the impact of carbapenemase resistance on mortality in patients infected with *Enterobacterales*. Except for one paper, in the bulk of the studies included in the meta-analysis, the risk ratio is more significant than unity, which shows a clear trend in the mechanism of resistance and mortality [137]. For *P. aeruginosa*, Yu Zhang et al. address the issue of mortality and carbapenem-resistance, finding a higher number of deaths in patients with carbapenem-resistant *P. aeruginosa* infections than in those without [138], and this phenomenon can be seen in any microorganism carbapenemase-producer.

Identifying the type of carbapenemase present is essential, as not all antibiotics can effectively inhibit it with a single molecule [139]. Clinicians need to choose specific antibiotics to overcome resistance mechanisms. A screening methodology is essential to determine whether carbapenemase production, efflux pumps, or lack of permeability are responsible for resistance to carbapenem.

IDENTIFICATION OF CARBAPENEMASES

The Hodge-modified method was previously used to identify carbapenemase in clinical strains, but it is unreliable due to the high rate of false negatives [140]. Phenotypic and molecular tests are the two ways to detect carbapenemase. After eliminating the Hodge-modified method, the modified Carbapenem Inactivation Method (CIM) was the best test. It is widely accepted and included in international guidelines, such as CLSI's M100 2023 [141, 142].

The test is easy to perform in a clinical microbiology lab. It only needs a tube with enriched broth and disposable, sterile bacteriological loops. For *Enterobacterales,* a 1μL loop is used, and a 10μL loop is used for *Pseudomonas aeruginosa*. A ten μg meropenem disk is placed and incubated for four hours at 37°C. Next, a 0.5 suspension of the McFarland scale is prepared with the pan-susceptible *E. coli* ATCC 25922 strain, incubated, and the test is read and interpreted after 18-24 hours.

The fact that two types of inoculants are used has to do with the growth kinetics of each microorganism; from the validation work, it was observed that it is enough to increase the inoculum to improve the sensitivity and specificity of the test. Likewise, the halos for the interpretation of the test are defined. 6-15 mm is considered a positive test; the problem strain is a carbapenemase producer because the problem strain had the selection pressure of the antibiotic and began to produce the enzyme that will degrade the antibiotic contained in the disk. Tests with halos 16-18 are indeterminate, which means that the test must be retested. However, if satellite colonies were found within the inhibition halo, the test would be positive, characteristic of OXA. Tests with > 19 mm are reported as unfavorable, meaning that the strains are not producers [141, 142]. This test has been shown to have good levels of sensitivity and specificity > 99% for KPC, NDM, VIM, IMP, IMI, SPM, SME, and OXA-type among *Enterobacterales* and > 97% sensitivity > 97% specificity of 100% for detection of KPC, NDM, VIM, IMP, IMI, SPM, and OXA-type carbapenemase among *P. aeruginosa* (Fig. **5**) [141].

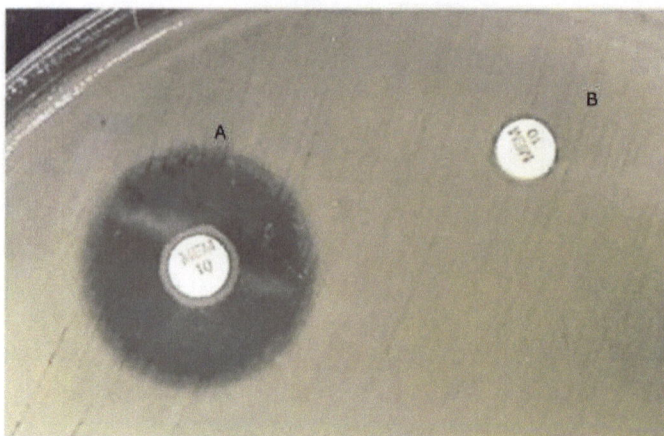

Fig. (5). A) Negative test from a non-producer of carbapenemase; B) positive test from a carbapenemase producer able to degrade the antibiotic in the disk. Image taken from CLSI M100, 2024 [141].

Virginia M. Pierce et al. established the parameters for determining whether a test is positive, negative, or intermediate in their work [143]. Initially, this test was only validated for *Enterobacterales*, using a 1 µL inoculation loop to define the inoculum. The next step was to demonstrate using the carbapenem inactivation method in *P. aeruginosa* and *A. baumannii*, as these two microorganisms are commonly found in hospitals. Patricia J. Simner conducted multicenter research to evaluate the performance of Carba NP and the modified carbapenem inactivation method against *P. aeruginosa* and *A. baumannii*. The researchers standardized the inoculum using 1 µL and 10 µL inoculation loops. With the 1 µL inoculation loop, they found a sensitivity of 80% (CI 61.4-92.3%), but when they used the 10 µL loop, sensitivity increased to 100%, with specificity remaining the same in both cases (93.3%). However, the test did not perform well for *A. baumannii*, with a sensitivity of only 34.8% (CI 21.4-50.3) using the 1 µL loop. When a 10 µL inoculation was used to standardize, the sensitivity improved substantially, rising to 98.32%. Although this test showed good percentages of sensitivity and specificity for *Enterobacterales* and *Pseudomonas aeruginosa*, it was not very successful for *A. baumannii,* possibly due to the mechanism of resistance OXA type [142]. The Hodge-modified method was later replaced, and the mCIM was introduced in CLSI guidelines as the screening method for carbapenemase [141], which became one of the most widely used methods worldwide. With the elimination of the modified Hodge method, clinical microbiology laboratories had a new tool that could detect carbapenemase with high sensitivity and specificity, with high clinical implications. The next step was to improve and develop a method to screen carbapenemases according to their hydrolytic nature, distinguishing between serine ß-lactamases and metallo ß-lactamases. This improvement was achieved by testing concentrations of chelators, such as EDTA, to remove the hydrolytic cofactor, and thus detect the presence of a metallo-β-lactamase [141]. Concentrations of 100 μM and five mM were tested, with the latter being the ideal concentration for 100% sensitivity and specificity [143]. This methodology was included again in CLSI M100 as a tool that clinical microbiology laboratories can use to detect metallo-β-lactamases [141].

Another method included in CLSI M100 to detect carbapenemase is the one developed by Dr. Laurent Poirel and Patrice Nordmann, which is faster than mCIM due to its colorimetric detection. This method was described before mCIM in 2015. Rapidec Carba NP (also known as Carba NP for Nordmann and Poirel) is based on carbapenem hydrolysis, specifically imipenem. The first step involves lysing the bacteria with a buffer and adding 100 μL of the lysate to a tube containing imipenem + cilastatin + phenol red as an indicator. If the strain being tested is a carbapenemase producer, it can disrupt the ß-lactam ring of carbapenem, changing the indicator color from red to yellow. Some enzymes can hydrolyze the carbapenem in at least 15 minutes, but the test must be read within 2 hours [144]. This test has been validated for *Enterobacterales* and *P. aeruginosa* (Fig. **6**).

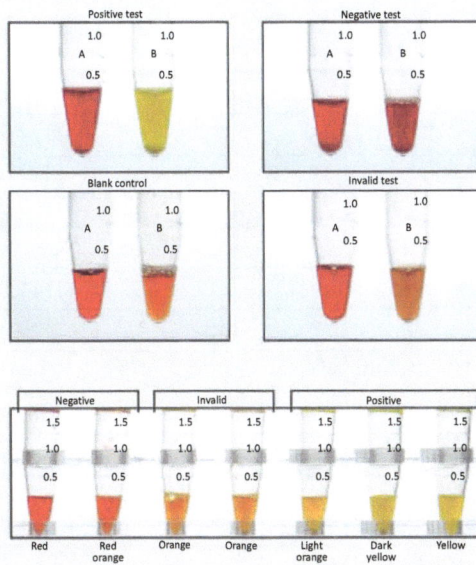

Fig. (6). Color interpretation. Image taken from the CLSI M100 [141].

If time is the variable to consider, then the ideal test is the lateral flow, also known as the immunochromatographic test. This test is one of the fastest because the carbapenemase type can be determined in only 15 minutes.

The relevance of epidemiological information is centered on the fact that it is necessary not only to make decisions but also to think of strategies to detect this resistance mechanism. In this way, Latania K. Logan and Robert A. Weinstein were responsible for defining the distribution of carbapenemase around the world. Their paper, published in 2017, identified five enzymes widely distributed around the globe. KPC, VIM, NDM, IMP, and OXA-48 were the information used to develop a new strategy known as Carba5 [145], created by a French group. They designed the test, which became popular, showing a sensitivity of 100% (according to the authors) and a specificity of 95% [146]. This method has been widely tested and is well-defined in terms of the proteins that can be detected. To realize the test, taking at least three strains and resuspending them in a tube with five drops of lysis buffer is necessary. After the sample is added to the test area of the cartridge, a waiting time of 15 minutes is required. The presence of carbapenemase is observable, with a red line in the letter corresponding to each enzyme [146]. The detection limit is too low for each protein. The lowest concentration detected is for NDM (150 pg/mL), and the highest is for KPC (KPC 600 pg/mL) [147]. In Table **3**, enzymes detected with Carba 5 are shown.

Table 3. Enzymes detected with Carba 5.

Group of Microorganisms	Enzymes	Variants Detected with Carba 5
Enterobacterales	KPC	2, 3, 4, 5, 6, 7, 12, 14, 28, and 39
	OXA-48 like	48, 162, 163, 181, 204, 232, 244, 245, 370, 405, 436, 484, 505, 515, 517, 519, 535, and 793
	VIM	1, 2, 4, 5, 6, 11, 19, 23, 27, and 31
	IMP	1, 4, 6, 8, 10, 11, 14, 26, 47, and 58
	NDM	1, 3, 4, 5, 6, 7, 9, 11, and 19
Pseudomonas aeruginosa	KPC	2, 3, and 5
	OXA-48 like	181
	VIM	2, 5, and 11
	IMP	1, 2, 5, 7, 13, 14, 15, 16, 18, 19, 26, 29, 31, 37, 39, 46, 56, 63, 71, and 79
	NDM	1

Detecting carbapenemases in clinical microbiology labs is essential, as these resistance mechanisms can significantly impact patients' treatment plans. Various methods, including inhibitors and molecular biology, are available for detecting carbapenemases.

The inhibitor method involves using inhibitors, such as boronic acid or EDTA, to identify the type of carbapenem hydrolytic activity. The process consists of preparing an inoculum sample test of 0.5 McFarland Scale, streaking the inoculum onto a Mueller-Hinton plate, and placing a disk with several molecules onto the agar.

One disk of inhibitor is used, followed by a meropenem disk alone. A positive sample is one where meropenem has an increased halo > 5 mm compared to the halo with meropenem alone. This test has a sensitivity of 92% and a specificity of 94% [148 - 151].

Molecular biology is a valuable tool in clinical microbiology labs, with commercial platforms offering > 99% sensitivity and 100% specificity. These platforms are used for the molecular search of carbapenemases in clinical samples, and most are included in syndromic molecular panels. However, some variants may not be included, such as GES with carbapenemase activity, which is common in clinical strains of *P. aeruginosa* and not detected by some techniques or molecular commercial kits [152 - 154]. Two commercial tools are Xpert Carba R of Cepheid and BioFire FilmArray; others are available. In-home tools are still the strategy used in many laboratories due to their adaptability and practically to amplify any target, and carbapenemase-coding genes are no exception (Table **4**) [155].

Table 4. Commercial molecular methods for carbapenemase detection.

Identification Strategy	Type of Carbapenemase Detected	Sensibility	Specificity	Reference
BioFire BCID2	KPC IMP NDM VIM OXA-48 like	94.9	99.7	[156]
BioFire Pneumonia	KPC IMP NDM VIM OXA-48 like	100	97	[157]
Xpert Carba R	KPC NDM VIM IMP-1 OXA (48, 181 y 232)	98	97	[158]

(Table 4) cont.....

Identification Strategy	Type of Carbapenemase Detected	Sensibility	Specificity	Reference
Entero-DR assay Seegene	KPC NDM VIM	100	92	[159]
BD-MAX CPO	KPC NDM VIM/IMP OXA-48 like	95.7	96.5	[160]

Finally, sequencing is another method for studying the genome of microorganisms and their resistance mechanisms. Although sequencing is mainly available in high-income countries, it helps detect carbapenemases. Regardless of the process, searching for carbapenemases in all clinical microbiology labs is essential to ensure appropriate patient treatment.

CONCLUSION

The final message is that we must have various robust methods to detect and identify carbapenemases, supporting physicians in the optimal prescription of antimicrobials according to guidelines, as well as promoting rational antimicrobial use practices.

ACKNOWLEDGEMENTS

The present work would not have been possible without the invaluable support of Ms. Graciela Flores and Ms. Rodrigo Martínez, thanks to their work and participation.

REFERENCES

[1]   Riley JC. Estimates of Regional and Global Life Expectancy, 1800–2001. Popul Dev Rev 2005; 31(3): 537-43.
[http://dx.doi.org/10.1111/j.1728-4457.2005.00083.x]

[2]   Hutchings MI, Truman AW, Wilkinson B. Antibiotics: past, present and future. Curr Opin Microbiol 2019; 51: 72-80.
[http://dx.doi.org/10.1016/j.mib.2019.10.008] [PMID: 31733401]

[3]   Abraham EP, Chain E, Fletcher CM, *et al.* Further Observations on Penicillin. Lancet 1941; 238(6155): 177-89.
[http://dx.doi.org/10.1016/S0140-6736(00)72122-2]

[4]   Lobanovska M, Pilla G. Penicillin's Discovery and Antibiotic Resistance: Lessons for the Future? Yale J Biol Med 2017; 90(1): 135-45.
[PMID: 28356901]

[5]   Tackling Drug-Resistant Infections Globally: Final Report and Recommendations. Review on Antimicrobial Resistance. Review on Antimicrobial Resistance 2016; pp. 1-80.

[6]   Review on Antibiotic resisitance Antimicrobial Resistance : Tackling a crisis for the health and wealth of nations. Review on Antimicrobial Resistance. Review on Antimicrobial Resistance 2014; pp. 1-16.

[7] Murray CJL, Ikuta KS, Sharara F, *et al.* Global burden of bacterial antimicrobial resistance in 2019: a systematic analysis. Lancet 2022; 399(10325): 629-55.
[http://dx.doi.org/10.1016/S0140-6736(21)02724-0] [PMID: 35065702]

[8] World Health Organization. WHO publishes list of bacteria for which new antibiotics are urgently needed 2017. https://www.who.int/news/item/27-02-2017-who-publishes-list-of-bacteria--or-which-new-antibiotics-are-urgently-needed

[9] World Health Organization. 2020. https://www.who.int/director-general/speeches/detail/who-direct-r-general-s-opening-remarks-at-the-media-briefing-on-covid-19---11-march-2020

[10] López-Jácome LE, Fernández-Rodríguez D, Franco-Cendejas R, *et al.* Increment Antimicrobial Resistance During the COVID-19 Pandemic: Results from the Invifar Network. Microb Drug Resist 2022; 28(3): 338-45.
[http://dx.doi.org/10.1089/mdr.2021.0231] [PMID: 34870473]

[11] Organization PAH. Antimicrobial Resistance, Fueled by the COVID-19 Pandemic. Policy Brief 2021 2022. https://iris.paho.org/handle/10665.2/55864

[12] CDC. 2021. https://www.cdc.gov/antimicrobial-resistance/data-research/threats/cov-d-19.html?CDC_AAref_Val=https://www.cdc.gov/drugresistance/covid19.html

[13] Zhanel GG, Wiebe R, Dilay L, *et al.* Comparative review of the carbapenems. Drugs 2007; 67(7): 1027-52.
[http://dx.doi.org/10.2165/00003495-200767070-00006] [PMID: 17488146]

[14] Papp-Wallace KM, Endimiani A, Taracila MA, Bonomo RA. Carbapenems: past, present, and future. Antimicrob Agents Chemother 2011; 55(11): 4943-60.
[http://dx.doi.org/10.1128/AAC.00296-11] [PMID: 21859938]

[15] Armstrong T, Fenn SJ, Hardie KR. JMM Profile: Carbapenems: a broad-spectrum antibiotic. J Med Microbiol 2021; 70(12): 001462.
[http://dx.doi.org/10.1099/jmm.0.001462] [PMID: 34889726]

[16] Kahan JS, Kahan FM, Goegelman R, *et al.* Thienamycin, a new. BETA.-lactam antibiotic. I. Discovery, taxonomy, isolation and physical properties. J Antibiot (Tokyo) 1979; 32(1): 1-12.
[http://dx.doi.org/10.7164/antibiotics.32.1] [PMID: 761989]

[17] Weaver SS, Bodey GP, LeBlanc BM. Thienamycin: new beta-lactam antibiotic with potent broad-spectrum activity. Antimicrob Agents Chemother 1979; 15(4): 518-21.
[http://dx.doi.org/10.1128/AAC.15.4.518] [PMID: 380462]

[18] Codjoe F, Donkor E. Carbapenem Resistance: A Review. Med Sci (Basel) 2017; 6(1): 1.
[http://dx.doi.org/10.3390/medsci6010001] [PMID: 29267233]

[19] Hawkey PM, Livermore DM. Carbapenem antibiotics for serious infections. BMJ 2012; 344(may31 1): e3236-6.
[http://dx.doi.org/10.1136/bmj.e3236] [PMID: 22654063]

[20] Pastel DA. Imipenem-cilastatin sodium, a broad-spectrum carbapenem antibiotic combination. Am J Health Syst Pharm 1986; 43(10): 2630-44.
[http://dx.doi.org/10.1093/ajhp/43.10.2630] [PMID: 3530614]

[21] Fukasawa M, Sumita Y, Harabe ET, *et al.* Stability of meropenem and effect of 1 beta-methyl substitution on its stability in the presence of renal dehydropeptidase I. Antimicrob Agents Chemother 1992; 36(7): 1577-9.
[http://dx.doi.org/10.1128/AAC.36.7.1577] [PMID: 1510457]

[22] Hashizume T, Ishino F, Nakagawa JI, Tamaki S, Matsuhashi M. Studies on the mechanism of action of imipenem (N-formimidoylthienamycin) *in vitro*: Binding to the penicillin-binding proteins (PBPs) in *Escherichia coli* and *Pseudomonas aeruginosa*, and inhibition of enzyme activities due to the PBPs in *E. coli*. J Antibiot (Tokyo) 1984; 37(4): 394-400.

[http://dx.doi.org/10.7164/antibiotics.37.394] [PMID: 6427167]

[23] Tipper DJ, Strominger JL. Mechanism of action of penicillins: a proposal based on their structural similarity to acyl-D-alanyl-D-alanine. Proc Natl Acad Sci USA 1965; 54(4): 1133-41.
[http://dx.doi.org/10.1073/pnas.54.4.1133] [PMID: 5219821]

[24] Sauvage E, Kerff F, Terrak M, Ayala JA, Charlier P. The penicillin-binding proteins: structure and role in peptidoglycan biosynthesis. FEMS Microbiol Rev 2008; 32(2): 234-58.
[http://dx.doi.org/10.1111/j.1574-6976.2008.00105.x] [PMID: 18266856]

[25] Ghuysen JM. Use of bacteriolytic enzymes in determination of wall structure and their role in cell metabolism. Bacteriol Rev 1968; 32(4_pt_2): 425-64.
[http://dx.doi.org/10.1128/br.32.4_pt_2.425-464.1968] [PMID: 4884715]

[26] Macheboeuf P, Contreras-Martel C, Job V, Dideberg O, Dessen A. Penicillin Binding Proteins: key players in bacterial cell cycle and drug resistance processes. FEMS Microbiol Rev 2006; 30(5): 673-91.
[http://dx.doi.org/10.1111/j.1574-6976.2006.00024.x] [PMID: 16911039]

[27] Goffin C, Ghuysen JM. Multimodular penicillin-binding proteins: an enigmatic family of orthologs and paralogs. Microbiol Mol Biol Rev 1998; 62(4): 1079-93.
[http://dx.doi.org/10.1128/MMBR.62.4.1079-1093.1998] [PMID: 9841666]

[28] Born P, Breukink E, Vollmer W. *In vitro* synthesis of cross-linked murein and its attachment to sacculi by PBP1A from *Escherichia coli*. J Biol Chem 2006; 281(37): 26985-93.
[http://dx.doi.org/10.1074/jbc.M604083200] [PMID: 16840781]

[29] Suzuki H, Nishimura Y, Hirota Y. On the process of cellular division in *Escherichia coli*: a series of mutants of *E. coli* altered in the penicillin-binding proteins. Proc Natl Acad Sci USA 1978; 75(2): 664-8.
[http://dx.doi.org/10.1073/pnas.75.2.664] [PMID: 345275]

[30] Denome SA, Elf PK, Henderson TA, Nelson DE, Young KD. *Escherichia coli* mutants lacking all possible combinations of eight penicillin binding proteins: viability, characteristics, and implications for peptidoglycan synthesis. J Bacteriol 1999; 181(13): 3981-93.
[http://dx.doi.org/10.1128/JB.181.13.3981-3993.1999] [PMID: 10383966]

[31] Meberg BM, Sailer FC, Nelson DE, Young KD. Reconstruction of *Escherichia coli mrcA* (PBP 1a) mutants lacking multiple combinations of penicillin binding proteins. J Bacteriol 2001; 183(20): 6148-9.
[http://dx.doi.org/10.1128/JB.183.20.6148-6149.2001] [PMID: 11567017]

[32] Schiffer G, Höltje JV. Cloning and characterization of PBP 1C, a third member of the multimodular class A penicillin-binding proteins of *Escherichia coli*. J Biol Chem 1999; 274(45): 32031-9.
[http://dx.doi.org/10.1074/jbc.274.45.32031] [PMID: 10542235]

[33] Ropp PA, Hu M, Olesky M, Nicholas RA. Mutations in *ponA*, the gene encoding penicillin-binding protein 1, and a novel locus, *penC*, are required for high-level chromosomally mediated penicillin resistance in *Neisseria gonorrhoeae*. Antimicrob Agents Chemother 2002; 46(3): 769-77.
[http://dx.doi.org/10.1128/AAC.46.3.769-777.2002] [PMID: 11850260]

[34] Den Blaauwen T, de Pedro MA, Nguyen-Distèche M, Ayala JA. Morphogenesis of rod-shaped sacculi. FEMS Microbiol Rev 2008; 32(2): 321-44.
[http://dx.doi.org/10.1111/j.1574-6976.2007.00090.x] [PMID: 18291013]

[35] Zapun A, Contreras-Martel C, Vernet T. Penicillin-binding proteins and β-lactam resistance. FEMS Microbiol Rev 2008; 32(2): 361-85.
[http://dx.doi.org/10.1111/j.1574-6976.2007.00095.x] [PMID: 18248419]

[36] Zapun A, Vernet T, Pinho MG. The different shapes of cocci. FEMS Microbiol Rev 2008; 32(2): 345-60.
[http://dx.doi.org/10.1111/j.1574-6976.2007.00098.x] [PMID: 18266741]

[37]   Braun V, Gnirke H, Henning U, Rehn K. Model for the structure of the shape-maintaining layer of the *Escherichia coli* cell envelope. J Bacteriol 1973; 114(3): 1264-70.
[http://dx.doi.org/10.1128/jb.114.3.1264-1270.1973] [PMID: 4576404]

[38]   Wientjes FB, Woldringh CL, Nanninga N. Amount of peptidoglycan in cell walls of gram-negative bacteria. J Bacteriol 1991; 173(23): 7684-91.
[http://dx.doi.org/10.1128/jb.173.23.7684-7691.1991] [PMID: 1938964]

[39]   Garde S, Chodisetti PK, Reddy M. Peptidoglycan: Structure, Synthesis, and Regulation. Ecosal Plus 2021; 9(2): ecosalplus.ESP-0010-2020.
[http://dx.doi.org/10.1128/ecosalplus.ESP-0010-2020] [PMID: 33470191]

[40]   Weidel W, Frank H, Martin HH. The rigid layer of the cell wall of *Escherichia coli* strain B. J Gen Microbiol 1960; 22(1): 158-66.
[http://dx.doi.org/10.1099/00221287-22-1-158] [PMID: 13843470]

[41]   Weidel W, Pelzer H. Bagshaped Macromolecules—A New Outlook on Bacterial Cell Walls. In: Purich D (Ed.) Advances in Enzymology - and Related Areas of Molecular Biology. John Wiley & Sons, Inc. 1964; 26.
[http://dx.doi.org/10.1002/9780470122716.ch5]

[42]   Work E. The mucopeptides of bacterial cell walls. A review. J Gen Microbiol 1961; 25(2): 167-89.
[http://dx.doi.org/10.1099/00221287-25-2-167] [PMID: 13786683]

[43]   Schleifer KH, Kandler O. Peptidoglycan types of bacterial cell walls and their taxonomic implications. Bacteriol Rev 1972; 36(4): 407-77.
[http://dx.doi.org/10.1128/br.36.4.407-477.1972] [PMID: 4568761]

[44]   van Heijenoort J. Lipid intermediates in the biosynthesis of bacterial peptidoglycan. Microbiol Mol Biol Rev 2007; 71(4): 620-35.
[http://dx.doi.org/10.1128/MMBR.00016-07] [PMID: 18063720]

[45]   Barreteau H, Kovač A, Boniface A, Sova M, Gobec S, Blanot D. Cytoplasmic steps of peptidoglycan biosynthesis. FEMS Microbiol Rev 2008; 32(2): 168-207.
[http://dx.doi.org/10.1111/j.1574-6976.2008.00104.x] [PMID: 18266853]

[46]   Badet B, Vermoote P, Haumont PY, Lederer F, Le Goffic F. Glucosamine synthetase from *Escherichia coli*: purification, properties, and glutamine-utilizing site location. Biochemistry 1987; 26(7): 1940-8.
[http://dx.doi.org/10.1021/bi00381a023] [PMID: 3297136]

[47]   Mengin-Lecreulx D, van Heijenoort J. Identification of the *glmU* gene encoding N-acetylglucosamin--1-phosphate uridyltransferase in *Escherichia coli*. J Bacteriol 1993; 175(19): 6150-7.
[http://dx.doi.org/10.1128/jb.175.19.6150-6157.1993] [PMID: 8407787]

[48]   Mengin-Lecreulx D, van Heijenoort J. Copurification of glucosamine-1-phosphate acetyltransferase and N-acetylglucosamine-1-phosphate uridyltransferase activities of *Escherichia coli*: characterization of the glmU gene product as a bifunctional enzyme catalyzing two subsequent steps in the pathway for UDP-N-acetylglucosamine synthesis. J Bacteriol 1994; 176(18): 5788-95.
[http://dx.doi.org/10.1128/jb.176.18.5788-5795.1994] [PMID: 8083170]

[49]   Mengin-Lecreulx D, van Heijenoort J. Characterization of the essential gene *glmM* encoding phosphoglucosamine mutase in *Escherichia coli*. J Biol Chem 1996; 271(1): 32-9.
[http://dx.doi.org/10.1074/jbc.271.1.32] [PMID: 8550580]

[50]   Marquardt JL, Siegele DA, Kolter R, Walsh CT. Cloning and sequencing of *Escherichia coli murZ* and purification of its product, a UDP-N-acetylglucosamine enolpyruvyl transferase. J Bacteriol 1992; 174(17): 5748-52.
[http://dx.doi.org/10.1128/jb.174.17.5748-5752.1992] [PMID: 1512209]

[51]   Benson TE, Marquardt JL, Marquardt AC, Etzkorn FA, Walsh CT. Overexpression, purification, and mechanistic study of UDP-N-acetylenolpyruvylglucosamine reductase. Biochemistry 1993; 32(8):

2024-30.
[http://dx.doi.org/10.1021/bi00059a019] [PMID: 8448160]

[52]   Kahan FM, Kahan JS, Cassidy PJ, Kropp H. The mechanism of action of fosfomycin (phosphonomycin). Ann N Y Acad Sci 1974; 235(1): 364-86.
[http://dx.doi.org/10.1111/j.1749-6632.1974.tb43277.x] [PMID: 4605290]

[53]   Liger D, Masson A, Blanot D, Van Heijenoort J, Parquet C. Over-production, purification and properties of the uridine-diphosphate-N-acetylmuramate:L-alanine ligase from *Escherichia coli*. Eur J Biochem 1995; 230(1): 80-7.
[http://dx.doi.org/10.1111/j.1432-1033.1995.0080i.x] [PMID: 7601127]

[54]   Doublet P, van Heijenoort J, Bohin JP, Mengin-Lecreulx D. The *murI* gene of *Escherichia coli* is an essential gene that encodes a glutamate racemase activity. J Bacteriol 1993; 175(10): 2970-9.
[http://dx.doi.org/10.1128/jb.175.10.2970-2979.1993] [PMID: 8098327]

[55]   Pratviel-Sosa F, Mengin-Lecreulx D, van HEIJENOORT J. Over-production, purification and properties of the uridine diphosphate *N*-acetylmuramoyl-L-alanine: D-glutamate ligase from *Escherichia coli*. Eur J Biochem 1991; 202(3): 1169-76.
[http://dx.doi.org/10.1111/j.1432-1033.1991.tb16486.x] [PMID: 1765076]

[56]   Michaud C, Mengin-Lecreulx D, van HEIJENOORT J, Blanot D. Over☐production, purification and properties of the uridine☐diphosphate☐ *N* ☐acetylmuramoyl☐ L ☐alanyl☐ D ☐glutamate: *meso* ☐2,6☐diaminopimelate ligase from *Escherichia coli*. Eur J Biochem 1990; 194(3): 853-61.
[http://dx.doi.org/10.1111/j.1432-1033.1990.tb19479.x] [PMID: 2269304]

[57]   Duncan K, Van Heijenoort J, Walsh CT. Purification and characterization of the D-alanyl-D-alan-ne-adding enzyme from *Escherichia coli*. Biochemistry 1990; 29(9): 2379-86.
[http://dx.doi.org/10.1021/bi00461a023] [PMID: 2186811]

[58]   Wild J, Hennig J, Lobocka M, Walczak W, Kłopotowski T. Identification of the dadX gene coding for the predominant isozyme of alanine racemase in *Escherichia coli* K12. Mol Gen Genet 1985; 198(2): 315-22.
[http://dx.doi.org/10.1007/BF00383013] [PMID: 3920477]

[59]   Wijsman HJW. The characterization of an alanine racemase mutant of *Escherichia coli*. Genet Res 1972; 20(3): 269-77.
[http://dx.doi.org/10.1017/S001667230001380X] [PMID: 4594607]

[60]   Zawadzke LE, Bugg TDH, Walsh CT. Existence of two D-alanine:D-alanine ligases in *Escherichia coli*: cloning and sequencing of the ddlA gene and purification and characterization of the DdlA and DdlB enzymes. Biochemistry 1991; 30(6): 1673-82.
[http://dx.doi.org/10.1021/bi00220a033] [PMID: 1993184]

[61]   Neuhaus FC, Lynch JL. The Enzymatic Synthesis of D-Alanyl-D-alanine. III. On the Inhibition of D-Alanyl-D-alanine Synthetase by the Antibiotic D-Cycloserine *. Biochemistry 1964; 3(4): 471-80.
[http://dx.doi.org/10.1021/bi00892a001] [PMID: 14188160]

[62]   Mengin-Lecreulx D, van Heijenoort J, Park JT. Identification of the mpl gene encoding UDP- N-acetylmuramate: L-alanyl-gamma-D-glutamyl-meso-diaminopimelate ligase in *Escherichia coli* and its role in recycling of cell wall peptidoglycan. J Bacteriol 1996; 178(18): 5347-52.
[http://dx.doi.org/10.1128/jb.178.18.5347-5352.1996] [PMID: 8808921]

[63]   Ikeda M, Wachi M, Jung HK, Ishino F, Matsuhashi M. The *Escherichia coli mraY* gene encoding UDP-N-acetylmuramoyl-pentapeptide: undecaprenyl-phosphate   phospho-N-acetylmuramoyl-pentapeptide transferase. J Bacteriol 1991; 173(3): 1021-6.
[http://dx.doi.org/10.1128/jb.173.3.1021-1026.1991] [PMID: 1846850]

[64]   Ghachi ME, Bouhss A, Blanot D, Mengin-Lecreulx D. The *bacA* gene of *Escherichia coli* encodes an undecaprenyl pyrophosphate phosphatase activity. J Biol Chem 2004; 279(29): 30106-13.
[http://dx.doi.org/10.1074/jbc.M401701200] [PMID: 15138271]

[65]   Ghachi ME, Derbise A, Bouhss A, Mengin-Lecreulx D. Identification of multiple genes encoding membrane proteins with undecaprenyl pyrophosphate phosphatase (UppP) activity in *Escherichia coli*. J Biol Chem 2005; 280(19): 18689-95.
[http://dx.doi.org/10.1074/jbc.M412277200] [PMID: 15778224]

[66]   TouzÉ T, Mengin-Lecreulx D. Undecaprenyl Phosphate Synthesis. Ecosal Plus 2008; 3(1): 10.1128/ecosalplus.4.7.1.7.
[http://dx.doi.org/10.1128/ecosalplus.4.7.1.7] [PMID: 26443724]

[67]   Stone KJ, Strominger JL. Mechanism of action of bacitracin: complexation with metal ion and C $_{55}$ - isoprenyl pyrophosphate. Proc Natl Acad Sci USA 1971; 68(12): 3223-7.
[http://dx.doi.org/10.1073/pnas.68.12.3223] [PMID: 4332017]

[68]   Kumar S, Rubino FA, Mendoza AG, Ruiz N. The bacterial lipid II flippase MurJ functions by an alternating-access mechanism. J Biol Chem 2019; 294(3): 981-90.
[http://dx.doi.org/10.1074/jbc.RA118.006099] [PMID: 30482840]

[69]   Kuk ACY, Hao A, Guan Z, Lee SY. Visualizing conformation transitions of the Lipid II flippase MurJ. Nat Commun 2019; 10(1): 1736.
[http://dx.doi.org/10.1038/s41467-019-09658-0] [PMID: 30988294]

[70]   Hvorup RN, Winnen B, Chang AB, Jiang Y, Zhou XF, Saier MH Jr. The multidrug/oligosaccharidyl-lipid/polysaccharide (MOP) exporter superfamily. Eur J Biochem 2003; 270(5): 799-813.
[http://dx.doi.org/10.1046/j.1432-1033.2003.03418.x] [PMID: 12603313]

[71]   Ruiz N. Lipid Flippases for Bacterial Peptidoglycan Biosynthesis. Lipid Insights 2018; 8(s1)2015; : 21-31.
[http://dx.doi.org/10.4137/Lpi.s31783]

[72]   Liu X, Meiresonne NY, Bouhss A, den Blaauwen T. FtsW activity and lipid II synthesis are required for recruitment of MurJ to midcell during cell division in *Escherichia coli*. Mol Microbiol 2018; 109(6): 855-84.
[http://dx.doi.org/10.1111/mmi.14104] [PMID: 30112777]

[73]   Taguchi A, Welsh MA, Marmont LS, *et al.* FtsW is a peptidoglycan polymerase that is functional only in complex with its cognate penicillin-binding protein. Nat Microbiol 2019; 4(4): 587-94.
[http://dx.doi.org/10.1038/s41564-018-0345-x] [PMID: 30692671]

[74]   Reichmann NT, Tavares AC, Saraiva BM, *et al.* SEDS–bPBP pairs direct lateral and septal peptidoglycan synthesis in Staphylococcus aureus. Nat Microbiol 2019; 4(8): 1368-77.
[http://dx.doi.org/10.1038/s41564-019-0437-2] [PMID: 31086309]

[75]   Welsh MA, Schaefer K, Taguchi A, Kahne D, Walker S. Direction of chain growth and substrate preferences of shape, elongation, division, and sporulation-family peptidoglycan glycosyltransferases. J Am Chem Soc 2019; 141(33): 12994-7.
[http://dx.doi.org/10.1021/jacs.9b06358] [PMID: 31386359]

[76]   Stoker NG, Pratt JM, Spratt BG. Identification of the rodA gene product of *Escherichia coli*. J Bacteriol 1983; 155(2): 854-9.
[http://dx.doi.org/10.1128/jb.155.2.854-859.1983] [PMID: 6348029]

[77]   Matsuzawa H, Hayakawa K, Sato T, Imahori K. Characterization and genetic analysis of a mutant of *Escherichia coli* K-12 with rounded morphology. J Bacteriol 1973; 115(1): 436-42.
[http://dx.doi.org/10.1128/jb.115.1.436-442.1973] [PMID: 4577747]

[78]   Ishino F, Park W, Tomioka S, *et al.* Peptidoglycan synthetic activities in membranes of *Escherichia coli* caused by overproduction of penicillin-binding protein 2 and rodA protein. J Biol Chem 1986; 261(15): 7024-31.
[http://dx.doi.org/10.1016/S0021-9258(19)62717-1] [PMID: 3009484]

[79]   Meeske AJ, Riley EP, Robins WP, *et al.* SEDS proteins are a widespread family of bacterial cell wall polymerases. Nature 2016; 537(7622): 634-8.

[http://dx.doi.org/10.1038/nature19331] [PMID: 27525505]

[80]   Emami K, Guyet A, Kawai Y, *et al.* RodA as the missing glycosyltransferase in *Bacillus subtilis* and antibiotic discovery for the peptidoglycan polymerase pathway. Nat Microbiol 2017; 2(3): 16253.
[http://dx.doi.org/10.1038/nmicrobiol.2016.253] [PMID: 28085152]

[81]   Henriques AO, Glaser P, Piggot PJ, Moran CP Jr. Control of cell shape and elongation by the *rodA* gene in *Bacillus subtilis*. Mol Microbiol 1998; 28(2): 235-47.
[http://dx.doi.org/10.1046/j.1365-2958.1998.00766.x] [PMID: 9622350]

[82]   Perkins HR. Specificity of combination between mucopeptide precursors and vancomycin or ristocetin. Biochem J 1969; 111(2): 195-205.
[http://dx.doi.org/10.1042/bj1110195] [PMID: 5763787]

[83]   Sjodt M, Brock K, Dobihal G, *et al.* Structure of the peptidoglycan polymerase RodA resolved by evolutionary coupling analysis. Nature 2018; 556(7699): 118-21.
[http://dx.doi.org/10.1038/nature25985] [PMID: 29590088]

[84]   Sjodt M, Rohs PDA, Gilman MSA, *et al.* Structural coordination of polymerization and crosslinking by a SEDS–bPBP peptidoglycan synthase complex. Nat Microbiol 2020; 5(6): 813-20.
[http://dx.doi.org/10.1038/s41564-020-0687-z] [PMID: 32152588]

[85]   Cho H, Wivagg CN, Kapoor M, *et al.* Bacterial cell wall biogenesis is mediated by SEDS and PBP polymerase families functioning semi-autonomously. Nat Microbiol 2016; 1(10): 16172.
[http://dx.doi.org/10.1038/nmicrobiol.2016.172] [PMID: 27643381]

[86]   Doi M, Wachi M, Ishino F, *et al.* Determinations of the DNA sequence of the mreB gene and of the gene products of the mre region that function in formation of the rod shape of *Escherichia coli* cells. J Bacteriol 1988; 170(10): 4619-24.
[http://dx.doi.org/10.1128/jb.170.10.4619-4624.1988] [PMID: 3049542]

[87]   Jones LJF, Carballido-López R, Errington J. Control of cell shape in bacteria: helical, actin-like filaments in *Bacillus subtilis*. Cell 2001; 104(6): 913-22.
[http://dx.doi.org/10.1016/S0092-8674(01)00287-2] [PMID: 11290328]

[88]   Kruse T, Møller-Jensen J, Løbner-Olesen A, Gerdes K. Dysfunctional MreB inhibits chromosome segregation in *Escherichia coli*. EMBO J 2003; 22(19): 5283-92.
[http://dx.doi.org/10.1093/emboj/cdg504] [PMID: 14517265]

[89]   Shih YL, Rothfield L. The bacterial cytoskeleton. Microbiol Mol Biol Rev 2006; 70(3): 729-54.
[http://dx.doi.org/10.1128/MMBR.00017-06] [PMID: 16959967]

[90]   Wang S, Arellano-Santoyo H, Combs PA, Shaevitz JW. Actin-like cytoskeleton filaments contribute to cell mechanics in bacteria. Proc Natl Acad Sci USA 2010; 107(20): 9182-5.
[http://dx.doi.org/10.1073/pnas.0911517107] [PMID: 20439764]

[91]   van den Ent F, Amos LA, Löwe J. Prokaryotic origin of the actin cytoskeleton. Nature 2001; 413(6851): 39-44.
[http://dx.doi.org/10.1038/35092500] [PMID: 11544518]

[92]   Carballido-Lopez R. The actin-like MreB proteins in *Bacillus subtilis* a new turn. Front Biosci (Schol Ed) 2012; S4(4): 1582-606.
[http://dx.doi.org/10.2741/s354]

[93]   Hussain S, Wivagg CN, Szwedziak P, *et al.* MreB filaments align along greatest principal membrane curvature to orient cell wall synthesis. eLife 2018; 7: e32471.
[http://dx.doi.org/10.7554/eLife.32471] [PMID: 29469806]

[94]   Morgenstein RM, Bratton BP, Nguyen JP, Ouzounov N, Shaevitz JW, Gitai Z. RodZ links MreB to cell wall synthesis to mediate MreB rotation and robust morphogenesis. Proc Natl Acad Sci USA 2015; 112(40): 12510-5.
[http://dx.doi.org/10.1073/pnas.1509610112] [PMID: 26396257]

[95] Bendezú FO, Hale CA, Bernhardt TG, de Boer PAJ. RodZ (YfgA) is required for proper assembly of the MreB actin cytoskeleton and cell shape in *E. coli*. EMBO J 2009; 28(3): 193-204.
[http://dx.doi.org/10.1038/emboj.2008.264] [PMID: 19078962]

[96] Shiomi D, Sakai M, Niki H. Determination of bacterial rod shape by a novel cytoskeletal membrane protein. EMBO J 2008; 27(23): 3081-91.
[http://dx.doi.org/10.1038/emboj.2008.234] [PMID: 19008860]

[97] Alyahya SA, Alexander R, Costa T, Henriques AO, Emonet T, Jacobs-Wagner C. RodZ, a component of the bacterial core morphogenic apparatus. Proc Natl Acad Sci USA 2009; 106(4): 1239-44.
[http://dx.doi.org/10.1073/pnas.0810794106] [PMID: 19164570]

[98] van den Ent F, Johnson CM, Persons L, de Boer P, Löwe J. Bacterial actin MreB assembles in complex with cell shape protein RodZ. EMBO J 2010; 29(6): 1081-90.
[http://dx.doi.org/10.1038/emboj.2010.9] [PMID: 20168300]

[99] Bertsche U, Breukink E, Kast T, Vollmer W. *In vitro* murein peptidoglycan synthesis by dimers of the bifunctional transglycosylase-transpeptidase PBP1B from *Escherichia coli*. J Biol Chem 2005; 280(45): 38096-101.
[http://dx.doi.org/10.1074/jbc.M508646200] [PMID: 16154998]

[100] Nakagawa J, Tamaki S, Tomioka S, Matsuhashi M. Functional biosynthesis of cell wall peptidoglycan by polymorphic bifunctional polypeptides. Penicillin-binding protein 1Bs of *Escherichia coli* with activities of transglycosylase and transpeptidase. J Biol Chem 1984; 259(22): 13937-46.
[http://dx.doi.org/10.1016/S0021-9258(18)89835-0] [PMID: 6389538]

[101] Schwartz B, Markwalder JA, Wang Y, Lipid II. Lipid II: total synthesis of the bacterial cell wall precursor and utilization as a substrate for glycosyltransfer and transpeptidation by penicillin binding protein (PBP) 1b of *Escherichia coli*. J Am Chem Soc 2001; 123(47): 11638-43.
[http://dx.doi.org/10.1021/ja0166848] [PMID: 11716719]

[102] Typas A, Banzhaf M, Gross CA, Vollmer W. From the regulation of peptidoglycan synthesis to bacterial growth and morphology. Nat Rev Microbiol 2012; 10(2): 123-36.
[http://dx.doi.org/10.1038/nrmicro2677] [PMID: 22203377]

[103] Egan AJF, Errington J, Vollmer W. Regulation of peptidoglycan synthesis and remodelling. Nat Rev Microbiol 2020; 18(8): 446-60.
[http://dx.doi.org/10.1038/s41579-020-0366-3] [PMID: 32424210]

[104] Leclercq S, Derouaux A, Olatunji S, *et al.* Interplay between Penicillin-binding proteins and SEDS proteins promotes bacterial cell wall synthesis. Sci Rep 2017; 7(1): 43306.
[http://dx.doi.org/10.1038/srep43306] [PMID: 28233869]

[105] Müller P, Ewers C, Bertsche U, *et al.* The essential cell division protein FtsN interacts with the murein (peptidoglycan) synthase PBP1B in *Escherichia coli*. J Biol Chem 2007; 282(50): 36394-402.
[http://dx.doi.org/10.1074/jbc.M706390200] [PMID: 17938168]

[106] Mueller EA, Egan AJF, Breukink E, Vollmer W, Levin PA. Plasticity of *Escherichia coli* cell wall metabolism promotes fitness and antibiotic resistance across environmental conditions. eLife 2019; 8: e40754.
[http://dx.doi.org/10.7554/eLife.40754] [PMID: 30963998]

[107] Impey RE, Hawkins DA, Sutton JM, Soares da Costa TP. Overcoming Intrinsic and Acquired Resistance Mechanisms Associated with the Cell Wall of Gram-Negative Bacteria. Antibiotics (Basel) 2020; 9(9): 623.
[http://dx.doi.org/10.3390/antibiotics9090623] [PMID: 32961699]

[108] Reuland EA, Halaby T, Hays JP, *et al.* Plasmid-mediated AmpC: prevalence in community-acquired isolates in Amsterdam, the Netherlands, and risk factors for carriage. PLoS One 2015; 10(1): e0113033.
[http://dx.doi.org/10.1371/journal.pone.0113033] [PMID: 25587716]

[109] Sirot D. Extended-spectrum plasmid-mediated -lactamases. J Antimicrob Chemother 1995; 36 (Suppl. A): 19-34.
[http://dx.doi.org/10.1093/jac/36.suppl_A.19] [PMID: 8543494]

[110] Kopotsa K, Osei Sekyere J, Mbelle NM. Plasmid evolution in carbapenemase-producing *Enterobacteriaceae*: A review. Ann N Y Acad Sci 2019; 1457(1): 61-91.
[http://dx.doi.org/10.1111/nyas.14223] [PMID: 31469443]

[111] Doi Y, Wachino J, Arakawa Y. Aminoglycoside Resistance. Infect Dis Clin North Am 2016; 30(2): 523-37.
[http://dx.doi.org/10.1016/j.idc.2016.02.011] [PMID: 27208771]

[112] Leclercq R. Mechanisms of resistance to macrolides and lincosamides: nature of the resistance elements and their clinical implications. Clin Infect Dis 2002; 34(4): 482-92.
[http://dx.doi.org/10.1086/324626] [PMID: 11797175]

[113] Iqbal MS, Rahman M, Islam R, *et al.* Plasmid-mediated sulfamethoxazole resistance encoded by the sul2 gene in the multidrug-resistant *Shigella flexneri* 2a isolated from patients with acute diarrhea in Dhaka, Bangladesh. PLoS One 2014; 9(1): e85338.
[http://dx.doi.org/10.1371/journal.pone.0085338] [PMID: 24416393]

[114] Dong N, Zeng Y, Wang Y, *et al.* Distribution and spread of the mobilised RND efflux pump gene cluster tmexCD-toprJ in clinical Gram-negative bacteria: a molecular epidemiological study. Lancet Microbe 2022; 3(11): e846-56.
[http://dx.doi.org/10.1016/S2666-5247(22)00221-X] [PMID: 36202114]

[115] Sundström L, Rådström P, Swedberg G, Sköld O. Site-specific recombination promotes linkage between trimethoprim- and sulfonamide resistance genes. Sequence characterization of dhfrV and sulI and a recombination active locus of Tn21. Mol Gen Genet 1988; 213(2-3): 191-201.
[http://dx.doi.org/10.1007/BF00339581] [PMID: 3054482]

[116] Cambray G, Guerout AM, Mazel D. Integrons. Annu Rev Genet 2010; 44(1): 141-66.
[http://dx.doi.org/10.1146/annurev-genet-102209-163504] [PMID: 20707672]

[117] Lim HM, Pène JJ, Shaw RW. Cloning, nucleotide sequence, and expression of the *Bacillus cereus* 5/B/6 beta-lactamase II structural gene. J Bacteriol 1988; 170(6): 2873-8.
[http://dx.doi.org/10.1128/jb.170.6.2873-2878.1988] [PMID: 3131315]

[118] Podglajen I, Breuil J, Bordon F, Gutmann L, Collatz E. A silent carbapenemase gene in strains of *Bacteroides fragilis* can be expressed after a one-step mutation. FEMS Microbiol Lett 1992; 91(1): 21-30.
[http://dx.doi.org/10.1111/j.1574-6968.1992.tb05178.x] [PMID: 1577251]

[119] Massidda O, Rossolini GM, Satta G. The *Aeromonas hydrophila cphA* gene: molecular heterogeneity among class B metallo-beta-lactamases. J Bacteriol 1991; 173(15): 4611-7.
[http://dx.doi.org/10.1128/jb.173.15.4611-4617.1991] [PMID: 1856163]

[120] Naas T, Nordmann P. Analysis of a carbapenem-hydrolyzing class A beta-lactamase from Enterobacter cloacae and of its LysR-type regulatory protein. Proc Natl Acad Sci USA 1994; 91(16): 7693-7.
[http://dx.doi.org/10.1073/pnas.91.16.7693] [PMID: 8052644]

[121] Yigit H, Queenan AM, Anderson GJ, *et al.* Novel carbapenem-hydrolyzing β-lactamase, KPC-1, from a carbapenem-resistant strain of *Klebsiella pneumoniae*. Antimicrob Agents Chemother 2001; 45(4): 1151-61.
[http://dx.doi.org/10.1128/AAC.45.4.1151-1161.2001] [PMID: 11257029]

[122] Yigit H, Queenan AM, Rasheed JK, *et al.* Carbapenem-resistant strain of *Klebsiella oxytoca* harboring carbapenem-hydrolyzing β-lactamase KPC-2. Antimicrob Agents Chemother 2003; 47(12): 3881-9.
[http://dx.doi.org/10.1128/AAC.47.12.3881-3889.2003] [PMID: 14638498]

[123] Bush K, Jacoby GA, Medeiros AA. A functional classification scheme for beta-lactamases and its

correlation with molecular structure. Antimicrob Agents Chemother 1995; 39(6): 1211-33.
[http://dx.doi.org/10.1128/AAC.39.6.1211] [PMID: 7574506]

[124] Ambler RP. The structure of β-lactamases. Philos Trans R Soc Lond B Biol Sci 1980; 289(1036): 321-31.
[http://dx.doi.org/10.1098/rstb.1980.0049] [PMID: 6109327]

[125] Hall BG, Barlow M. Revised Ambler classification of β-lactamases. J Antimicrob Chemother 2005; 55(6): 1050-1.
[http://dx.doi.org/10.1093/jac/dki130] [PMID: 15872044]

[126] Hammoudi Halat D, Ayoub Moubareck C. The current burden of carbapenemases: Review of significant properties and dissemination among gram-negative bacteria. Antibiotics (Basel) 2020; 9(4): 186.
[http://dx.doi.org/10.3390/antibiotics9040186] [PMID: 32316342]

[127] Naas T, Oueslati S, Bonnin RA, et al. Beta-lactamase database (BLDB) – structure and function. J Enzyme Inhib Med Chem 2017; 32(1): 917-9.
[http://dx.doi.org/10.1080/14756366.2017.1344235] [PMID: 28719998]

[128] Hammoudi D, Ayoub Moubareck C, Karam Sarkis D. How to detect carbapenemase producers? A literature review of phenotypic and molecular methods. J Microbiol Methods 2014; 107: 106-18.
[http://dx.doi.org/10.1016/j.mimet.2014.09.009] [PMID: 25304059]

[129] Abboud MI, Damblon C, Brem J, et al. Interaction of Avibactam with Class B Metallo-β-Lactamases. Antimicrob Agents Chemother 2016; 60(10): 5655-62.
[http://dx.doi.org/10.1128/AAC.00897-16] [PMID: 27401561]

[130] Bahr G, González LJ, Vila AJ. Metallo-β-lactamases in the Age of Multidrug Resistance: From Structure and Mechanism to Evolution, Dissemination, and Inhibitor Design. Chem Rev 2021; 121(13): 7957-8094.
[http://dx.doi.org/10.1021/acs.chemrev.1c00138] [PMID: 34129337]

[131] Antunes N, Fisher J. Acquired Class D β-Lactamases. Antibiotics (Basel) 2014; 3(3): 398-434.
[http://dx.doi.org/10.3390/antibiotics3030398] [PMID: 27025753]

[132] Bush K, Jacoby GA. Updated functional classification of β-lactamases. Antimicrob Agents Chemother 2010; 54(3): 969-76.
[http://dx.doi.org/10.1128/AAC.01009-09] [PMID: 19995920]

[133] Turton JF, Woodford N, Glover J, Yarde S, Kaufmann ME, Pitt TL. Identification of *Acinetobacter baumannii* by detection of the $bla_{OXA-51-like}$ carbapenemase gene intrinsic to this species. J Clin Microbiol 2006; 44(8): 2974-6.
[http://dx.doi.org/10.1128/JCM.01021-06] [PMID: 16891520]

[134] CDC. https://arpsp.cdc.gov/profile/arln/crpa

[135] NCBI. https://www.ncbi.nlm.nih.gov/pathogens/refgene#KPC

[136] Yong D, Toleman MA, Giske CG, et al. Characterization of a new metallo-β-lactamase gene, $bla_{(NDM-1)}$, and a novel erythromycin esterase gene carried on a unique genetic structure in *Klebsiella pneumoniae* sequence type 14 from India. Antimicrob Agents Chemother 2009; 53(12): 5046-54.
[http://dx.doi.org/10.1128/AAC.00774-09] [PMID: 19770275]

[137] Zhou R, Fang X, Zhang J, et al. Impact of carbapenem resistance on mortality in patients infected with *Enterobacteriaceae* : a systematic review and meta-analysis. BMJ Open 2021; 11(12): e054971.
[http://dx.doi.org/10.1136/bmjopen-2021-054971] [PMID: 34907071]

[138] Zhang Y, Chen XL, Huang AW, et al. Mortality attributable to carbapenem-resistant *Pseudomonas aeruginosa* bacteremia: a meta-analysis of cohort studies. Emerg Microbes Infect 2016; 5(1): 1-6.
[http://dx.doi.org/10.1038/emi.2016.22] [PMID: 27004762]

[139] Lasarte-Monterrubio C, Fraile-Ribot PA, Vázquez-Ucha JC, et al. Activity of cefiderocol,

imipenem/relebactam, cefepime/taniborbactam and cefepime/zidebactam against ceftolozane/tazobactam- and ceftazidime/avibactam-resistant *Pseudomonas aeruginosa*. J Antimicrob Chemother 2022; 77(10): 2809-15.
[http://dx.doi.org/10.1093/jac/dkac241] [PMID: 35904000]

[140]   Pasteran F, Mendez T, Rapoport M, Guerriero L, Corso A. Controlling false-positive results obtained with the Hodge and Masuda assays for detection of class a carbapenemase in species of *enterobacteriaceae* by incorporating boronic Acid. J Clin Microbiol 2010; 48(4): 1323-32.
[http://dx.doi.org/10.1128/JCM.01771-09] [PMID: 20181912]

[141]   Clinical and Laboratory Standards Institute (CLSI). M100. Performance Standards for Antimicrobial Susceptibility Testing. USA: Clinical and Laboratory Standards Institute (CLSI); 2024.

[142]   Simner PJ, Johnson JK, Brasso WB, *et al.* Multicenter Evaluation of the Modified Carbapenem Inactivation Method and the Carba NP for Detection of Carbapenemase-Producing *Pseudomonas aeruginosa* and Acinetobacter baumannii. J Clin Microbiol 2018; 56(1): e01369-17.
[http://dx.doi.org/10.1128/JCM.01369-17] [PMID: 29118172]

[143]   Sfeir MM, Hayden JA, Fauntleroy KA, *et al.* EDTA-Modified Carbapenem Inactivation Method: a Phenotypic Method for Detecting Metallo-β-Lactamase-Producing *Enterobacteriaceae*. J Clin Microbiol 2019; 57(5): e01757-18.
[http://dx.doi.org/10.1128/JCM.01757-18] [PMID: 30867235]

[144]   Poirel L, Nordmann P. Rapidec Carba NP Test for Rapid Detection of Carbapenemase Producers. J Clin Microbiol 2015; 53(9): 3003-8.
[http://dx.doi.org/10.1128/JCM.00977-15] [PMID: 26085619]

[145]   Logan LK, Weinstein RA. The epidemiology of carbapenem-resistant enterobacteriaceae: The impact and evolution of a global menace. J Infect Dis 2017; 215 (Suppl. 1): S28-36.
[http://dx.doi.org/10.1093/infdis/jiw282] [PMID: 28375512]

[146]   Hopkins KL, Meunier D, Naas T, Volland H, Woodford N. Evaluation of the NG-Test CARBA 5 multiplex immunochromatographic assay for the detection of KPC, OXA-48-like, NDM, VIM and IMP carbapenemases. J Antimicrob Chemother 2018; 73(12): 3523-6.
[http://dx.doi.org/10.1093/jac/dky342] [PMID: 30189008]

[147]   Carba 5 test - carbapenemase resistance detection -. [cited 24 Apr 2025]. Available: https://www.ngbiotech.com/ng-test-carba-5/#1658126971787-2015d01a-c08e

[148]   Mendez-Sotelo BJ, López-Jácome LE, Colín-Castro CA, *et al.* Comparison of lateral flow immunochromatography and phenotypic assays to PCR for the detection of carbapenemase-producing gram-negative bacteria, a multicenter experience in Mexico. Antibiotics (Basel) 2023; 12(1): 96.
[http://dx.doi.org/10.3390/antibiotics12010096] [PMID: 36671297]

[149]   Tsakris A, Kristo I, Poulou A, *et al.* Evaluation of boronic acid disk tests for differentiating KPC-possessing *Klebsiella pneumoniae* isolates in the clinical laboratory. J Clin Microbiol 2009; 47(2): 362-7.
[http://dx.doi.org/10.1128/JCM.01922-08] [PMID: 19073868]

[150]   Yong D, Lee K, Yum JH, Shin HB, Rossolini GM, Chong Y. Imipenem-EDTA disk method for differentiation of metallo-β-lactamase-producing clinical isolates of *Pseudomonas* spp. and *Acinetobacter* spp. J Clin Microbiol 2002; 40(10): 3798-801.
[http://dx.doi.org/10.1128/JCM.40.10.3798-3801.2002] [PMID: 12354884]

[151]   García-Betancur JC, Appel TM, Esparza G, *et al.* Update on the epidemiology of carbapenemases in Latin America and the Caribbean. Expert Rev Anti Infect Ther 2021; 19(2): 197-213.
[http://dx.doi.org/10.1080/14787210.2020.1813023] [PMID: 32813566]

[152]   El Sherif HM, Elsayed M, El-Ansary MR, Aboshanab KM, El Borhamy MI, Elsayed KM. BioFire filmArray BCID2 versus VITEK-2 system in determining microbial etiology and antibiotic-resistant genes of pathogens recovered from central line-associated bloodstream infections. Biology (Basel) 2022; 11(11): 1573.

[http://dx.doi.org/10.3390/biology11111573] [PMID: 36358274]

[153] Jin S, Lee JY, Park JY, Jeon MJ. Xpert Carba-R assay for detection of carbapenemase-producing organisms in patients admitted to emergency rooms. Medicine (Baltimore) 2020; 99(50): e23410.
[http://dx.doi.org/10.1097/MD.0000000000023410] [PMID: 33327265]

[154] Garza-González E, Bocanegra-Ibarias P, Bobadilla-del-Valle M, *et al.* Drug resistance phenotypes and genotypes in Mexico in representative gram-negative species: Results from the infivar network. PLoS One 2021; 16(3): e0248614.
[http://dx.doi.org/10.1371/journal.pone.0248614] [PMID: 33730101]

[155] Rojas-Larios F, Martínez-Guerra BA, López-Jácome LE, *et al.* Active surveillance of antimicrobial resistance and carbapenemase-encoding genes according to sites of care and age groups in Mexico: Results from the INVIFAR network. Pathogens 2023; 12(9): 1144.
[http://dx.doi.org/10.3390/pathogens12091144] [PMID: 37764952]

[156] Peri AM, Ling W, Furuya-Kanamori L, Harris PNA, Paterson DL. Performance of BioFire Blood Culture Identification 2 Panel (BCID2) for the detection of bloodstream pathogens and their associated resistance markers: a systematic review and meta-analysis of diagnostic test accuracy studies. BMC Infect Dis 2022; 22(1): 794.
[http://dx.doi.org/10.1186/s12879-022-07772-x] [PMID: 36266641]

[157] Kamel NA, Alshahrani MY, Aboshanab KM, El Borhamy MI. Evaluation of the bioFire filmArray pneumonia panel plus to the conventional diagnostic methods in determining the microbiological etiology of hospital-acquired pneumonia. Biology (Basel) 2022; 11(3): 377.
[http://dx.doi.org/10.3390/biology11030377] [PMID: 35336751]

[158] Duze ST, Thomas T, Pelego T, Jallow S, Perovic O, Duse A. Evaluation of Xpert Carba-R for detecting carbapenemase-producing organisms in South Africa. Afr J Lab Med 2023; 12(1): 1898.
[http://dx.doi.org/10.4102/ajlm.v12i1.1898] [PMID: 36756217]

[159] Mojica MF, De La Cadena E, Correa A, Appel TM, Pallares CJ, Villegas MV. Evaluation of Allplex™ Entero-DR assay for detection of antimicrobial resistance determinants from bacterial cultures. BMC Res Notes 2020; 13(1): 154.
[http://dx.doi.org/10.1186/s13104-020-04997-4] [PMID: 32178721]

[160] Yoo IY, Shin DP, Heo W, Ha SI, Cha YJ, Park YJ. Comparison of BD MAX Check-Points CPO assay with Cepheid Xpert Carba-R assay for the detection of carbapenemase-producing Enterobacteriaceae directly from rectal swabs. Diagn Microbiol Infect Dis 2022; 103(3): 115716.
[http://dx.doi.org/10.1016/j.diagmicrobio.2022.115716] [PMID: 35596981]

CHAPTER 5

Post-antibiotic and Resistance Era

Mariano Martínez Vázquez[1,*]

[1] *Department of Natural Products, Chemistry Institute University National Autonomous of México*

Abstract: The Centers for Disease Control and Prevention (CDC-USA) has categorized bacterial infections as urgent, serious, and concerning threats. In this chapter, we will discuss the antibiotics suggested by the CDC-USA to combat the urgent threats caused by drug-resistant bacteria, including carbapenem-resistant Acinetobacter and Enterobacteriaceae, *Candida auris*, Clostridioides difficile, and Drug-resistant Neisseria gonorrhoeae.

The chapter reviews the efficacy of carbapenem antibiotics when combined with β-lactamase inhibitors. Additionally, it discusses using quorum-sensing inhibitors to prevent virulence factors, focusing on halogen furanone-type compounds. Interestingly, such inhibitors are effective *in vitro,* but none are currently being evaluated in clinical settings. Finally, some syntheses of nanoparticles to counteract drug-resistant bacteria are reported. Nanoparticles made from metals, especially silver, and those synthesized using biodegradable polymeric materials have shown promising results in cytotoxic activity. However, further studies are needed to determine their effectiveness and toxicity.

Keywords: Bactericidal-drugs combination, Bactericidal nanoparticles, CDU-USA classification, Hybrids, Quorum Sensing System.

INTRODUCTION

Recent research has shed light on microorganisms' pivotal role in the human body throughout their lifespan. The human microbiome comprises various organisms, including bacteria, fungi, viruses, and other microbes. Its functions include aiding food digestion, vitamin synthesis, regulating the immune system, and protecting against pathogenic agents. Traditional research in human microbiology focused on identifying individual microbes, such as bacteria, fungi, and viruses, primarily from clinical isolates. However, contemporary molecular and biochemical analyses such as genomics, transcriptomics, proteomics, and metabolomics have

---

[*] **Corresponding author Mariano Martínez Vázquez:** Department of Natural Products, Chemistry Institute University National Autonomous of México; E-mail: marvaz@unam.mx

Mariano Martínez-Vázquez (Ed.)

facilitated the detection and classification of diverse microorganisms in different body parts, such as the gastrointestinal tract, skin, airway system, and urogenital tract.

These analyses have demonstrated that everyone possesses microbiota that contributes to health maintenance and disease prevention. This progressive understanding has significantly advanced our comprehension of the pathogenesis of various human diseases. These insights are expected to pave the way for developing diagnostic, therapeutic, and preventive measures in personalized/precision medicine [1].

Disrupting the balance of the microbiome can result in bacterial infections. These infections are caused by harmful bacteria invading and multiplying in the body, leading to symptoms like fever, pain, inflammation, and organ damage.

Bacterial infections can result in various illnesses, including strep throat, meningitis, sepsis, and pneumonia. They could also induce food poisoning. The microorganisms responsible for some of these diseases include *Streptococcus* spp., *Staphylococcus aureus*, *Escherichia coli* (*E. coli*), and *Salmonella* spp.

Maintaining a healthy microbiome through good hygiene practices, a balanced diet, and judicious use of antibiotics is crucial to mitigate the risk of bacterial infections. The human microbiota consists of bacteria, viruses, archaea, molds, yeasts, and protozoa. It resides in the skin, mucous membranes of cavities, and secretory glands. The symbiotic relationship between the human microbiota and the host helps modulate the host's physiological development and immune functions throughout life.

However, throughout history, there have been instances where new microorganisms have infected humans, causing widespread death and destruction. The Black Death is an example of such an infection, which ravaged Eurasia from 1347 to 1351, resulting in the deaths of millions. *Yersinia pestis* is believed to have caused Justinian's Plague in the 6th century and killed millions of people in the Byzantine Empire [2].

Conversely, it is generally assumed that the number of microorganisms in the human body is much greater than that of human cells. However, in 2016, a study was published in which, using new calculation techniques, it was concluded that there is an almost 1:1 relationship between the number of human cells and microorganisms [3].

A balance between human cells and microorganisms is essential for proper growth and development. When this equilibrium is broken, it can lead to health complications. The balance could be restored by antibiotic drugs, reducing the number of microorganisms causing the infection. However, these drugs are non-selective and can lead to complications such as the formation of kidney stones when taking sulfonamides, abnormal blood clotting when taking some cephalosporins, sensitivity to sunlight when taking tetracyclines, blood disorders when taking trimethoprim and deafness when taking erythromycin and aminoglycosides. Some people, especially older adults, may experience intestinal inflammation, which can cause severe bloody diarrhea [4].

Therefore, it is crucial to exercise caution while using, such drugs to prevent adverse effects and ensure the well-being of individuals. Recent research on the human microbiome has demonstrated that even healthy individuals display significant variations in microorganisms in areas such as the skin, gut, and vagina. While the factors contributing to this diversity are not yet fully understood, diet, environment, genetics, and initial exposure to microorganisms have been identified as possible influencers. The Human Microbiome Project examined many people and studied microbial communities in or on the human body. This project showed that each habitat has a diverse range of microbes that vary among healthy individuals. It was also found that each person has unique microbial niches within their body, and the diversity and abundance of these microbes can differ significantly from person to person. The study estimated that the healthy Western microbiome contains a great diversity of genera, enzyme families, and community configurations [5].

Antibiotics have long been used to treat bacterial infections. However, the emergence of antibiotic-resistant strains has made it difficult to eradicate these infections. Thus, the prevalence of resistant strains in clinical settings has made it challenging to eliminate bacterial infections.

Unquestionably, the evolution of drug-resistant bacteria was due to selection pressure, which increases an organism's ability to survive and reproduce. Organisms with advantageous traits are more likely to pass them on to future generations, leading to their prevalence in subsequent generations. For that, the microorganisms that are best suited to new conditions are the ones that have the highest likelihood of survival and reproduction, and their characteristics become increasingly common in future generations.

Antibiotic resistance (ABR) in microorganisms is caused by the selection pressure they face when exposed to different drugs. In the past, antibiotic resistance was limited to a specific place where a microbial strain developed resistance to a particular type of drug. However, the widespread and inappropriate use of antibiotics in humans and animals has led to multiple microorganisms being threatened by various antibiotics. These practices have contributed to the global spread of ABR, which is a complex issue that poses a significant threat to the health and well-being of humans and animals. Factors contributing to ABR include poor regulation of antibiotics (access and quality) in many parts of the world, increasing global demand for animal-origin food, human and domestic animal populations, and globalization and international trading. Consequently, ABR is now a widespread threat to humans, animals, and the environment, with transmission occurring between these sectors *via* complex pathways [6, 7].

Bacteria can naturally develop resistance to antibiotics, whether innate or acquired. This resistance could be examined from different perspectives, including population, pharmacokinetics, molecular biology, pharmacodynamics, and the clinician's point of view.

Some microorganisms have natural resistance mechanisms that can hinder the effects of antimicrobial agents. This intrinsic resistance is a characteristic of the bacteria and can involve one or more mechanisms, such as chromosomal β-lactamases or natural barriers. Understanding the concept of inherent resistance is essential for microbiologists to identify bacteria accurately.

Medical practitioners should avoid prescribing antibiotics that are ineffective against certain bacteria due to intrinsic resistance. For instance, ampicillin, amoxicillin, and first-generation cephalosporins should not be prescribed for *Enterobacter* spp., *Citrobacter freundii*, *Morganella morganii*, *Providencia* spp., and *Serratia* spp. (Fig. **1**) [8].

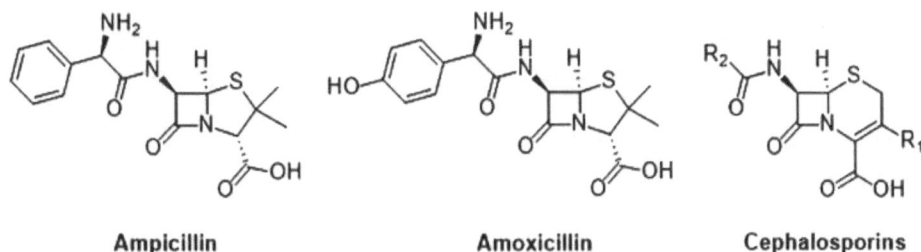

Ampicillin   **Amoxicillin**   **Cephalosporins**

Fig. (1). Chemical Structures of Ampicillin, Amoxicillin and Cephalosporins.

Intrinsic mechanisms refer to naturally occurring genes on the host's chromosome, such as AmpC β-lactamase of Gram-negative bacteria and many Multiple-Drugs Resistance (MDR) efflux systems. Understanding intrinsic resistance mechanisms is paramount in developing effective treatment strategies against bacterial infections.

The development of acquired resistance in bacteria is a notable genetic change that occurs when a bacterium becomes immune to a specific antimicrobial. This resistance can be temporary and is often influenced by environmental factors, as evidenced by genetic analysis. For example, when grown in anaerobic conditions, strains of *Escherichia coli* become resistant to aminoglycosides, while abundant magnesium ions in the environment confer resistance to polymyxin and aminoglycosides in *Pseudomonas aeruginosa*. Acquired resistance can also be permanent and is typically associated with extrachromosomal elements acquired from other mechanisms instead of chromosomal mutations.

Notably, some bacteria have developed resistance to drugs approved by the FDA within the past decade, indicating that the environment can serve as a source for both new and old resistance mechanisms. Furthermore, organisms exposed to the first commercially produced antibiotics have developed resistance to specific antibiotics. For instance, *Staphylococci* have developed penicillin resistance by making the penicillinase enzyme, which breaks down the antibiotics. Using new bactericidal drugs has created a continuous selection pressure, leading to new resistance mechanisms, such as new penicillin-binding proteins and enzymatic modification of drugs in MDR bacteria.

Bacterial resistance to antibiotics can arise from various mechanisms, including mutations in the target gene of the chromosome, acquisition of foreign genetic material, and the tolerance phenomenon. The latter should be regarded as a form of acquired resistance, even if the microorganism remains susceptible to medication.

Acquiring mobile genetic elements, including plasmids, is a particularly vital mechanism for bacterial resistance. Plasmids are circular DNA molecules that replicate independently of the chromosome and can transfer genetic information horizontally between bacteria through conjugation. In the evolution of bacterial antibiotic resistance, plasmids play a significant role in disseminating resistance genes among the most severe clinical pathogens.

Conjugative plasmids are the most significant contributors to the spread of antibiotic resistance in bacterial families such as Enterobacteriaceae and Enterococcaceae. Over time, some of these plasmid-bacterium associations have become remarkably successful, with specific Antibiotic Resistance (AR) plasmids strongly linked to bacterial lineages; for instance, associations between *Klebsiella pneumoniae* sequence type 11 (ST11) and plasmid pOXA-48 (carrying carbapenemase gene blaOXA-48) [9].

Nonetheless, the success of a plasmid-bacterium association depends on several key requirements. The fate of the plasmid in the microbial residents is determined by a combination of factors, including the intensity of its cost and mutations that compensate for this cost, the strength and frequency of selection for plasmid-encoded traits (*e.g.*, antibiotic resistance, colonization, virulence), the rate of plasmid loss due to segregation during bacterial division, the rate of horizontal plasmid transfer among bacteria through conjugation, and potential founder effects acting on a plasmid-bacterium association.

The evolutionary dynamics of plasmid-carrying bacteria will be a crucial determinant of the rise and spread of associations between antibiotic-resistant plasmids and bacterial pathogens *in vivo* [10]. Therefore, it is essential to understand the mechanisms of plasmid acquisition, maintenance, and dissemination to develop strategies to prevent and control the spread of antibiotic resistance.

CARBAPENEM ANTIBIOTICS

Carbapenem antibiotics are a potent class of antibiotics that can effectively treat severe bacterial infections caused by multidrug-resistant bacteria (MDR). Typically reserved as a last resort when other antibiotics have failed, carbapenems exhibit a structure like penicillin and cephalosporin but with a broader range of activity and better resistance to bacterial enzymes. They function by inhibiting the formation of the bacterial cell wall, ultimately resulting in bacterial death.

Examples of carbapenem antibiotics include penams, penems, cephems, oxapenams, monobactams, carbapenems, imipenems, meropenems, ertapenems, and doripenems.

Carbapenem antibiotics are used to treat severe bacterial infections resistant to other drugs. However, they are not recommended as the first option because they might promote the development of antibiotic-resistant bacteria. Therefore, it is crucial to use microbial susceptibility testing and follow best clinical practices before starting carbapenem therapy. Although all carbapenems have a standard β-lactam ring structure, their side chains can alter their spectrum of activity and pharmacokinetic properties (Fig. **2**).

As bacteria grow, their cell walls undergo continuous remodeling. Old walls break down, and new ones are formed with the help of various enzymes. These enzymes belong to a group of acyl serine transferases that play a vital role in the polymerization and cross-linking of peptidoglycan glycan strands. Carbapenem-type antibiotics could inhibit this remodeling.

Fig. (2). Chemical Structures of Penam, Penem, Cephem, Oxapenam, Monobactam, Carbapenam, Carbapenem, Imipenem, Meropenam, Ertapenam and Doripenam.

Classification of Carbapenem-type Antibiotics

Penams are a class of antibiotics that contain a thiazolidine ring fused to a β-lactam ring, with various side chains attached to the thiazolidine ring. It is effective against a narrow spectrum of Gram-positive bacteria.

Penems are antibiotics with a β-lactam ring like penams but with a carbon atom instead of a sulfur atom in the thiazolidine ring. The most active penem is imipenem, which showed activity against Gram-positive and Gram-negative bacteria.

Cephems are antibiotics with a six-member dihydrothiazine ring fused to a β-lactam ring. They have different side chains attached to the dihydrothiazine ring. Cephalexin is an antibiotic effective against many bacteria, including Gram-positive and Gram-negative strains. On the other hand, oxazepam possesses a fused 1,3-oxazolidine ring to a ring β-lactam system.

Monobactams are a class of antibiotics containing a monocyclic β-lactam ring structure with a side chain attached to the ring. Aztreonam, the most well-known monobactam, has a narrow spectrum of activity against Gram-negative bacteria.

Carbapenems are a class of antibiotics considered some of the most potent available. They are often used as a last resort treatment for serious bacterial infections, such as those caused by multidrug-resistant bacteria.

Activity of Imipenem, Meropenem, Ertapenem and Doripenem

Imipenem is a semisynthetic thienamycin that has demonstrated antibacterial activity. This broad-spectrum antibiotic is stable against a wide range of β-lactamases, making it a reliable option for treating bacterial infections. Imipenem is often co-administered with cilastatin, and the FDA recently approved a novel triple-drug product containing imipenem, cilastatin, and relebactam. In November 1985, the FDA first approved imipenem as the combination product Primaxin, marketed by Merck & Co [11].

Meropenem, a similar compound to imipenem, is known for its efficacy against Gram-negative bacteria. Meropenem is an antibacterial agent in the carbapenem-carboxylic acid class. It contains an azetidine ring and a pyrroline ring, which are substituted with 1-hydroxymethyl and 5-(dimethyl carbamoyl) pyrrolidine---altho, respectively. In addition to its therapeutic benefits, this drug can cause drug allergies [12].

β-Lactam antibiotics, such as penem-type, prevent the transpeptidase enzymes, also known as penicillin-binding proteins (PBPs), from accessing their natural substrates by blocking their active site. As a result, the cross-linking of the cell wall is prevented, and the cell dies due to osmotic instability of ions [13].

Ertapenem exhibits a longer half-life owing to its increased binding to plasma proteins. It is a type of broad-spectrum β-lactam antibiotic with bactericidal properties. Ertapenem is a 1-β-methyl carbapenem that binds to penicillin-binding proteins (PBPs), particularly PBPs 2 and 3, located on the bacterial cell wall. It inhibits the final transpeptidation step in synthesizing peptidoglycan, an essential bacterial cell wall component. This drug weakens and destroys the cell walls of bacteria, leading to their death. It is effective against anaerobic and both Gram-negative and Gram-positive aerobic bacteria. Additionally, it is stable against hydrolysis by various β-lactamases, including penicillinases, cephalosporinases, and extended-spectrum β-lactamases [14].

Doripenem displays broad-spectrum activity and bactericidal properties. It is resistant to β-lactamase and binds with penicillin-binding proteins (PBPs) found on the bacterial cell wall, particularly on PBPs 2 and 3. Doripenem weakens and eventually lyses the bacterial cell wall by inhibiting the final transpeptidation step responsible for synthesizing peptidoglycan, an essential component of the bacterial cell wall. According to studies, doripenem is 2- to 16-fold more potent than imipenem and demonstrates comparable efficacy to ertapenem and meropenem [15].

As mentioned, the cell walls continually remodel during bacterial growth. This process involves a group of enzymes integral to the polymerization and cross-linking of peptidoglycan glycan strands. The enzymes include transglycosylase, transpeptidase, carboxypeptidase, and endopeptidase.

These enzymes belong to a group of acyl serine transferases exhibiting a natural affinity for D-alanyl-D-alanine. This compound shares structural similarities with beta-lactam antibiotics, as depicted in Fig. (**3**). However, this structural resemblance can lead to confusion on the enzyme's part. Consequently, they can bind with β-lactam antibiotics in their active sites, disrupting their natural processes.

These antibiotics are usually given through injections or intravenous infusion since they are poorly absorbed orally. They can cure many bacterial infections, such as pneumonia, sepsis, urinary tract infections, and intra-abdominal infections.

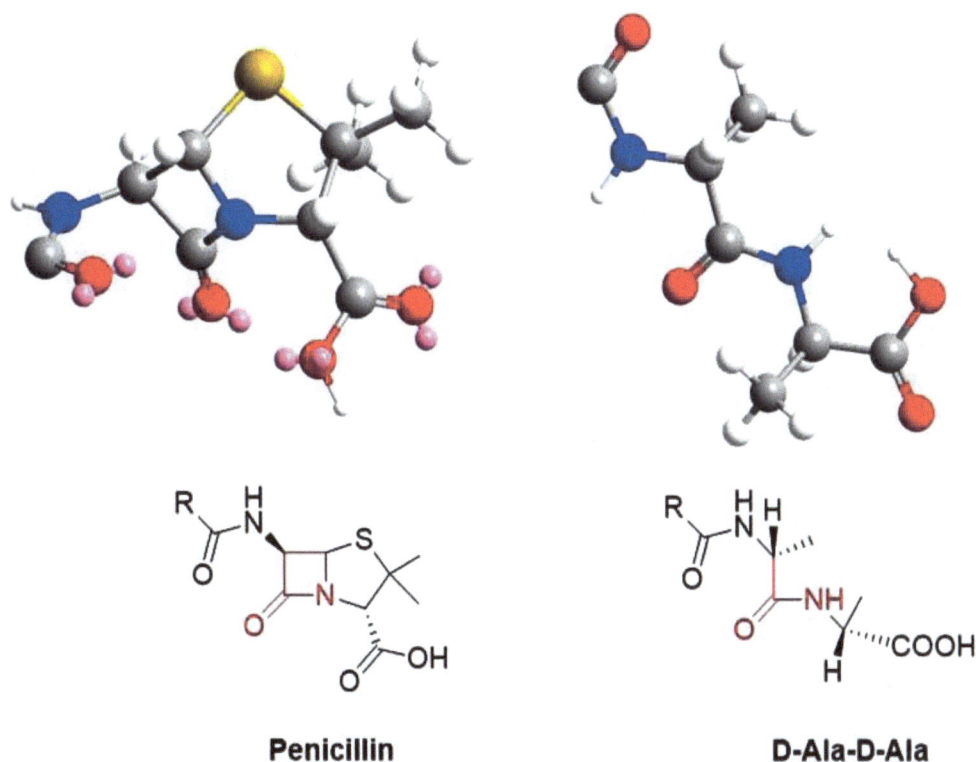

Penicillin   **D-Ala-D-Ala**

Fig. (3). Stereochemistry of Penicillin and D-Ala-D-Ala derivatives.

Resistance to Carbapenem Antibiotics

Carbapenem-resistant microorganisms (CRMOs) have become a significant concern for public health as they resist carbapenem antibiotics and can be challenging to treat. It is crucial to use these antibiotics sensibly and take measures to prevent the spread of CRMOs. Of particular concern are carbapenem-resistant Enterobacteriaceae (CRE), a cluster of bacteria that exhibit resistance to carbapenems and can result in serious infections such as pneumonia, bloodstream infections, and urinary tract infections. Preventing the proliferation of CRE and other carbapenemase-producing bacteria is a significant public health priority that necessitates infection control and antibiotic stewardship.

CRMOs can develop resistance to various classes of antibiotics through different mechanisms, such as changes in the bacterial cell wall or efflux pumps that remove antibiotics from the bacterial cell before they can take effect. The emergence of CRMOs is a significant threat to public health as it can make it challenging to treat severe bacterial infections and increase the risk of mortality. Therefore, implementing strategies such as infection control measures, surveillance, and the appropriate use of antibiotics is essential to prevent the spread of CRMOs.

Carbapenemases

Over the past ten years, there has been a notable increase in carbapenem-resistant Gram-negative bacteria. The resistance to carbapenems in many of these isolates is caused by carbapenemases β-lactamases, which can inactivate carbapenems and other β-lactam antibiotics. Presently, these enzymes are primarily categorized into three different β-lactamase classes, namely class A, B, and D. In Enterobacteriaceae, the most clinically significant carbapenemases are class A (enzymes of KPC-type), New Delhi metallo-β-lactamases (NDM), imipenemases (IMP), Verona integron-encoded metallo-β-lactamases (VIM), OXA-type carbapenemases, and class D carbapenemases of the OXA-48 type expressed by plasmids [16].

Class B MBLs, including NDM, VIM, and IMP, are zinc-dependent and require metal ions for their catalytic activity. Structural comparisons between KPC enzymes and the well-characterized TEM-1 and SHV-1 penicillinases have shown that the catalytic residues in KPCs are positioned to accommodate the bulky side chains of carbapenems, enabling efficient acylation and deacylation [17]. The genes encoding these carbapenemases are frequently found on plasmids, small pieces of DNA that can be easily transferred between bacteria. This implies that carbapenem resistance can rapidly spread between bacterial species, including those ordinarily susceptible to these antibiotics.

INFECTIONS

Classification CDC-USA

Considering the rise of infections and deaths induced by microorganisms, the Center for Disease Control and Prevention (CDC-USA) classified the infections as urgent, serious, and concerning threats. It is worth noting that the CDC list differs from that proposed by the WHO in 2017.

Urgent Threats

- Carbapenem-resistant *Acinetobacter.*
- *Candida auris.*
- *Clostridioides difficile.*
- Carbapenem-resistant Enterobacteriaceae (CRE).
- Drug-resistant *Neisseria gonorrhoeae.*

Serious Threats.

- Drug-resistant *Campylobacter.*
- Drug-resistant *Candida.*
- Extended-spectrum beta-lactamase (ESBL)-producing Enterobacteriaceae.
- Vancomycin-resistant *Enterococci* (VRE).
- Multidrug-resistant *Pseudomonas aeruginosa.*
- Drug-resistant *nontyphoidal Salmonella.*
- Drug-resistant *Salmonella* (Typhi serotype).
- Drug-resistant *Shigella.*
- Methicillin-resistant *Staphylococcus aureus* (MRSA).
- Drug-resistant *Streptococcus pneumoniae.*
- Drug-resistant *Tuberculosis.*

Concerning Threats

- Erythromycin-resistant group *A Streptococcus.*
- Clindamycin-resistant group *B Streptococcus.*
- Azole-resistant *Aspergillus fumigatus.*
- Drug-resistant *Mycoplasma genitalium.*
- Drug-resistant *Bordetella pertussis.*

Urgent Threats

Considering its lethality and ease of transmission, from the CDC list, in this section, we will only discuss Carbapenem-resistant *Acinetobacter*, Carbapenem-resistant Enterobacteriaceae (CRE*), Candida auris* (*C. auris*), *Clostridioides difficile* (*C. difficile), Drug-resistant Neisseria gonorrhoeae* (*N. gonorrhea*), and Multidrug-resistant *Pseudomonas aeruginosa.*

The evolution of *K. pneumoniae* to acquire resistance to various antibiotics illustrates, in general, the stages a species goes through to reach MDR or Xtreme Drug Resistant (XDR) strains. These species are also information reservoirs of potent β-lactamases, which can be distributed to other strains *via* transmissible plasmids [18].

Carbapenem-resistant Acinetobacter

Acinetobacter was once considered a harmless bacterium that occasionally caused infections. However, with the rise of mechanical ventilation, catheterization, and antibiotic use, *Acinetobacter* infections have become more frequent and severe.

Frequently, particular species of *Acinetobacter* are associated with hospital-acquired infections. *Acinetobacter's* ability to survive in dry environments and evade host defenses allows a high bacterial density and can lead to sepsis caused by lipopolysaccharide (LPS) mediated by receiver 4 (TLR4) [19].

In *A. baumannii*, the acquired resistance mechanisms are more frequent, mainly due to the production of different βlactamases. Class A (like TEM-1 and PSE-1), class B (metallo-β-lactamases (like blaNDM), class C (like *Acinetobacter*-derived cephalosporinases ADCs), and the most predominant class D (like blaOXA-58).

On the other hand, chromosomal or non-enzymatic mechanisms, such as membrane impermeability due to carbapenem-associated outer membrane protein (CarO) or overexpression of the AdeABC efflux pump, also contribute to antimicrobial resistance [20].

A possible way to eliminate opportunistic bacterial pathogens like *A. baumannii* is to repurpose already-used drugs. Anticancer drugs may be beneficial among these drugs due to their cytotoxic activities and similarities between bacterial infections and growing tumors. Then, cisplatin, mitomycin C (MMC) and melphalan, all DNA crosslinkers, and the metabolite 5-fluorouracil were tested for their effectiveness on the growth of *A. baumannii* ATCC BAA-747. The study showed that MMC was the most effective drug. At a concentration of 7 μg/mL, it inhibited 50% of growth, and at 25 μg/mL, it completely inhibited growth in the Luria-Bertani medium. Considering its activity, MMC was tested against a panel of 18 multidrug-resistant isolates, 3 sensitive only to colistin. MMC's minimum inhibitory and minimum bactericidal concentrations in all tested strains were similar to those of *A. baumannii* ATCC BAA-747. MMC could also effectively kill stationary-phase, persister, and biofilm cells. MMC eradicates actively growing bacterial and stationary-phase, persister, and biofilm cells. It can also enhance the survival rate of insect larvae *Galleria mellonella* infected with an otherwise lethal strain of *A. baumannii*. In the case of the antibiotic-sensitive *A. baumannii* ATCC BAA-747 strain and the multidrug-resistant (MDR) strains A560 and A578, the survival rate was increased from 0% to ≥53%. This suggests that MMC can potentially combat the emergent opportunistic pathogen *A. baumannii* [21].

The EURECA study, conducted between May 2016 and November 2018, involved 29 sites in 10 European countries. The study analyzed whole-genome and core-genome MLST (multilocus sequence typing) of *A. baumannii* isolates, revealing a high diversity among the strains. The most common sequence type was ST2, which accounted for 67.7% of the isolates (n=153). This *A. baumannii* strain collection from the EURECA study represents a unique and diverse repository of carbapenem-resistant isolates, which adds to the knowledge of the epidemiology and resistance genes harbored by these strains [22].

Carbapenem-resistant *Acinetobacter Baumannii* (CRAB) is a bacterium resistant to carbapenem antibiotics, typically employed to combat bacterial infections. This strain of bacteria is a significant public health concern as it can cause severe infections, including pneumonia, bloodstream infections, and wound infections, that are notoriously difficult to treat [23]. Notably, the optimal treatment for CRAB infections is contingent upon the specific strain of bacteria and the severity of the infection. Given its high antibiotic resistance rate, CRAB has been associated with a 70% mortality rate from infections caused by extensively drug-resistant (XDR) strains. Consequently, there is an urgent need for novel preventive and therapeutic options for *Acinetobacter* spp [24].

Several drugs treat CRAB infections, including Colistin, tigecycline, polymyxin B, sulbactam/ampicillin combination, cefiderocol, eravacycline, and plazomicin.

Colistin: This antibiotic is usually used as a last resort for treating CRAB infections because it can cause side effects like kidney damage. Colistin activity is due to the interaction and disruption of the outer membrane of the bacteria. Previously, colistin was discouraged due to its nephrotoxicity and neurotoxicity, and less toxic antibiotics like aminoglycosides were preferred. However, recent studies have shown that the incidence and severity of toxicity caused by colistin are lower than previously reported, and it has proven to be effective in treating MDR bacteria. Better formulations of colistimethate sodium, careful dosing, and critical care services have contributed to the safer use of colistin.

Unfortunately, *A. baumannii* colistin resistance has been described worldwide. In general, rates of colistin resistance have been increasing globally, especially in healthcare settings where antibiotic use is frequent and there is a higher risk of acquiring antibiotic-resistant infections. According to a 2020 global surveillance report by the World Health Organization (WHO), colistin resistance has been reported in all regions, ranging from less than 1% to more than 50% in some countries. The highest colistin resistance rates have been reported in countries with high antibiotic use, poor infection prevention, lack of control measures, and limited access to alternative treatments. However, even in countries with lower colistin resistance rates, there is a concern about the emergence and spread of resistant strains (WHO Surveillance Report 2020).

The exact resistance mechanism is unknown, but some studies suggest it may be associated with losing LPS or the PmrAB two-component system. It has been found that colistin monotherapy is ineffective in preventing resistance. Therefore, combination therapy is considered the best strategy against colistin-resistant *A.*

baumannii. The combinations of colistin/rifampicin and colistin/carbapenem have been extensively studied, and they have shown promising results both *in vitro* and *in vivo*, as well as in clinical trials. New peptides with good activity against colistin-resistant *A. baumannii* are currently under investigation [25, 26].

Tigecycline: This is the first member of the glycylcycline class (a 9---butylglycylamido derivative of minocycline). This broad-spectrum antibiotic treats various bacterial infections, including skin and soft tissue infections, intra-abdominal infections, and community-acquired pneumonia. Tigecycline works by inhibiting bacterial protein synthesis, which causes the elimination of bacteria. It is effective against Gram-positive and Gram-negative bacteria, including some strains resistant to other commonly used antibiotics. Tigecycline is also effective against CRAB (carbapenem-resistant *Acinetobacter* baumannii) and is often used with other antibiotics.

Tigecycline is administered intravenously at a treatment dosage depending on the type and severity of the infection. It is typically used as a last resort when other antibiotics have failed or are ineffective. Although tigecycline is generally well-tolerated, it may present effects such as nausea, vomiting, diarrhea, and headache. Additionally, it may reduce the effectiveness of oral contraceptives, which is why additional methods of birth control may be necessary while using tigecycline. Since tigecycline can treat infections caused by multidrug-resistant bacteria, it is a crucial antibiotic in managing severe infections, particularly in hospitals [27, 28].

Tigecycline has been used to treat multidrug-resistant *Acinetobacter baumannii* (MDR-AB) infections. However, there is an ongoing debate on its effectiveness against these ailments [29]. A systematic review and meta-analysis were conducted to evaluate the efficiency and security of tigecycline in healing MDR-AB infections. The study found no significant difference in mortality rates and clinical response between patients treated with tigecycline and those in the control group. Furthermore, subgroup analysis showed that in-hospital mortality rates were higher in patients treated with tigecycline. Tigecycline was also less effective in eradicating microbes, resulting in more extended hospital stays than the control group. Combination therapy using tigecycline did not affect mortality rates, clinical response, or microbiological response. Nonetheless, the patients tolerated tigecycline well; during treatment, the resistance emergence and superinfection rates were 12.47% and 19.11%, respectively. Based on these findings, a tigecycline-based regimen may not be the best option for treating MDR-AB infections [29].

Polymyxin B: Although discovered in 1947, the clinical use of polymyxins was brief due to concerns about potential kidney and nerve damage and the availability of safer alternatives. However, due to increased drug-resistant Gram-negative pathogens, polymyxins are now being used again as a last-resort drug. Unfortunately, their ability to cause kidney damage has limited their effectiveness. As a result, there is a need for new and improved polymyxin drugs that are safer and more effective. Various international organizations, including academic and industry groups, have been working to develop such drugs in recent years. Despite these efforts, there is still a need for a new and improved polymyxin drug for clinical use [30].

It's concerning that the number of reports of polymyxin resistance in Gram-negative bacteria has increased rapidly over the past decade. As a result, there is an urgent need to investigate the mechanism of polymyxin resistance and develop new polymyxin antibiotics to deal with this challenging situation. The primary cause of polymyxin resistance in Gram-negative bacteria is modifications to the lipopolysaccharide (LPS) molecule, a significant component of the bacterial outer membrane (OM). In most polymyxin-resistant strains, cationic moieties such as 4-amino-4-deoxy-L-arabinose (L-Ara4N), phosphoethanolamine (pEtN), or galactosamine are added to the lipid A or core components of LPS, which are believed to reduce the electrostatic interactions with polymyxins [31].

Various two-component systems (TCSs), such as PhoPQ and PmrAB, modulate the expression of most genes in the LPS modification pathway. Stimuli, including high Fe^{3+}, low pH, and antibiotics, are known to induce the expression of several genes (*e.g.*, eptA and arnT), leading to the translation of enzyme machinery that mediates the modification of LPS molecules with the cationic moieties [32].

The novel polymyxins MRX-8 and comparators were evaluated *in vitro* for potency and spectrum in a nonclinical study versus a considerable set of Gram-negative clinical isolates obtained in the USA from 2017 to 2020. *Enterobacterales*, *Pseudomonas aeruginosa*, and *A. baumannii* exhibit nearly identical antimicrobial susceptibility to MRX-8, colistin, and polymyxin B. All three polymyxin-class compounds did not show activity versus isolates presenting acquired or intrinsic resistance to polymyxins but retained activity against meropenem-resistant and multidrug-resistant isolate subsets. Due to that, MRX-8 reduced toxicity *in vitro*, and animal assays are now being evaluated in an initial clinical phase [30].

Sulbactam/ampicillin: This is a combination antibiotic used to treat bacterial infections caused by certain strains of bacteria. These strains produce β-lactamase enzymes that can break down and inactivate some antibiotics. Studies have shown that the sulbactam/ampicillin combination can be effective against certain strains of CREA. However, the effectiveness of this combination can vary depending on each bacterium and the individual patient. In some cases, other antibiotics or combination therapies may be more effective against CREA. For instance, a 35-year-old Japanese woman developed skin and smooth tissue disease due to carbapenem-resistant *A. baumannii*, which was immune to antibiotics other than ampicillin–sulbactam and colistin. This result indicated drug antagonism due to carbapenemase production by OXA-23. A hybrid therapy of intravenous ampicillin–sulbactam and meropenem was established, demonstrating efficacy. No changes are observed in aspartate- or alanine aminotransferases, blood urea nitrogen, and serum creatinine during treatment. After three days of therapy initiation, carbapenem-resistant *A. baumannii* was no longer detected in the wound exudates.

Cefiderocol: This new cephalosporin acts like the rest of β-lactam by inhibiting the bacterial cell wall through its fixation to penicillin-binding proteins (PBP). It is considered a hybrid of Ceftazidime and Cefepime, with the difference that a catechol group has in position 3. The presence of the catechol residue is responsible for cefiderocol´s siderophore activity, allowing the formation of chelates with iron and conducting the accumulation of the antibiotic in the periplasmic space, where it will join PBP.

Effective antibiotics for carbapenem-resistant *Acinetobacter* infections are lacking despite a substantial unmet medical need. Cefiderocol (Fetroja) has demonstrated potent *in vitro* activity against multidrug-resistant (MDR) *Acinetobacter* and was recently approved to treat drug-resistant Gram-negative pathogens [33]. However, its *in vivo* efficacy in preclinical infection models of cefiderocol-susceptible *Acinetobacter baumannii* is highly variable. Therefore, further research is required to evaluate the therapeutic potential of cefiderocol for treating *Acinetobacter* infections [34].

Eravacycline: This is a broad-spectrum antibiotic in the tetracycline class. It was created by Tetraphase Pharmaceuticals and authorized by the FDA in 2018 to treat complicated intra-abdominal infections (cIAI) in adults [35]. Eravacycline works by inhibiting bacterial protein synthesis, preventing bacterial growth and multiplication. It showed activity against an extensive collection of bacteria, including Gram-positive and Gram-negative strains, as well as antibiotic-resistant strains such as MRSA and ESBL-producing Enterobacteriaceae [36].

The effectiveness of eravacycline-based combinations varies among different species of bacteria. For example, when combined with β-lactams or polymyxin B, eravacycline can produce synergistic effects against carbapenem-resistant Gram-negative bacteria. A combination of eravacycline and polymyxin B may be a viable treatment option for carbapenem-resistant *E. coli* and *K. pneumoniae*.

Additionally, eravacycline, when combined with ceftazidime or a carbapenem antimicrobial, may be a suitable treatment for carbapenem-resistant *A. baumannii*. (Fig. **4**) [37].

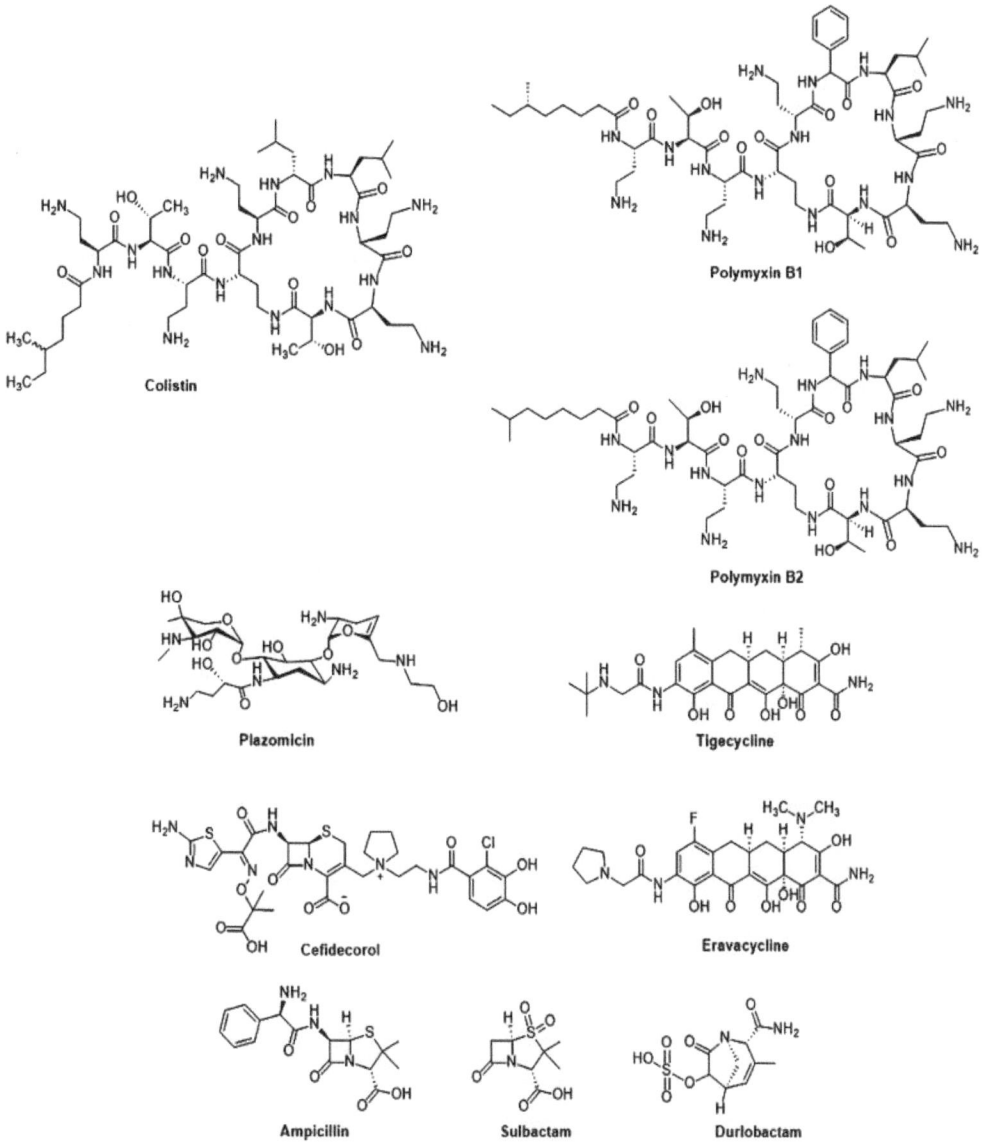

Fig. (4). Chemical structures of Colistin, Polymyxin B, Tigecycline, Plazomicin, Cefiderocol, Eravacycline, Ampicillin, Sulbactam, and Durlobactam.

According to reports, 50 patients were treated with eravacycline for infections during outpatient antibiotic therapy (OPAT). Of the 50 patients, 47 achieved clinical resolution while taking eravacycline. The drug was well tolerated, with nausea being the only adverse event reported, and none of the patients discontinued treatment. Due to single daily dosing, all patients received their eravacycline ending dose in the OPAT location or at home. *Clostridioides difficile* infection (CDI) is a problematic condition that often occurs after the use of broad-spectrum antibiotics. Interestingly, after the treatment with eravacycline, only one patient developed CDI 30 days post-treatment, despite many patients having reported a CDI in the 6 to 12 months before the treatment. Higher clinical achievement and microbiologic outcomes were shown by eravacycline treatment, consistent with the efficacy observed in clinical trials, including infections with ESBL- or KPC-producing isolates. Furthermore, the treatment did not produce many side effects in a challenging group of patients [38, 39].

Plazomicin: This is a type of drug known as an aminoglycoside. It works by binding to the bacterial 30S ribosomal subunit, which inhibits protein synthesis in a concentration-dependent way. This medication is primarily used to treat complicated urinary tract infections, such as pyelonephritis, caused by certain types of bacteria like those belonging to Enterobacteriaceae and *Pseudomonas aeruginosa*. It is designed to combat resistance mechanisms developed by these bacteria against other aminoglycoside antibiotics. Plazomicin is effective against a broad range of aerobic Gram-negative bacteria, including extended-spectrum β-lactamase-producing Enterobacteriaceae, carbapenem-resistant Enterobacteria ceae, carbapenem-resistant organisms, and those with aminoglycoside-modifying enzymes (Table **1**).

Table 1. Drugs against Carbapenem-resistant *Acinetobacter.*

Drugs in the clinic	Observations	Reference
Colistin	Colistin treats bacterial infections, particularly those caused by multidrug-resistant Gram-negative bacteria. It damages the bacterial cell membrane, which causes the bacteria to die. Due to its potential toxicity and side effects, colonistin is often used as a last-resort antibiotic when other treatments have failed.	[24]
Tigecycline	Tigecycline inhibits bacterial protein synthesis. It is often used to treat skin infections, intra-abdominal infections, and pneumonia, among other conditions.	[25]
Polymyxin B	Polymyxin B damages the bacterial cell membrane, causing the bacteria to die. This drug is particularly effective against Gram-negative bacteria, often resistant to other antibiotics.	[28]

(Table 1) cont.....

Drugs in the clinic	Observations	Reference
Sulbactam/ampicillin	A combination that could be effective against some strains of CRAB.	[23]
Rifampicin	Used in combination with other antibiotics to treat CRAB.	[23]
Cefiderocol	It acts both as a beta-lactam drug and a siderophore compound.	[23]
Drugs in Clinical Trials or recently approved		
Eravacycline	A tetracycline antibiotic that is effective against CRAB in clinical trials.	[36]
Plazomicin	An aminoglycoside antibiotic that is effective against CRAB *in vitro* and in clinical trials	[39]
Sulbactam-durlobactam	A combination of antibiotics that has shown promise in treating CRAB infections in clinical trials.	[177]

It is important to note that plazomicin does not work against all drug-resistant bacteria. Bacteria can become antibiotic-resistant in different ways, such as through mutations or acquiring resistance genes from other bacteria.

In phase III clinical trials, plazomicin is as effective as meropenem in treating complicated urinary tract infections (cUTIs). Plazomicin is also effective in diseases caused by carbapenem-resistant *Enterobacteriaceae*. Moreover, plazomicin was associated with a lower rate of mortality or significant disease-related complications (23.5% [4/17]) compared to colistin (50% [10/20]). However, patients who take plazomicin may experience some adverse reactions, such as hypo- or hypertension, decreased renal function, diarrhea, headache, nausea, and vomiting. Plazomicin, like other aminoglycosides, may cause problems related to the nervous and muscular systems, hearing loss, and harm to fetuses in pregnant women. Due to limited safety data, plazomicin is only recommended for treating complicated urinary tract infections (cUTIs) in adults when other treatment options are restricted or unavailable. Patients with kidney problems should have their dosage reduced and their drug levels monitored. Plazomicin is not recommended for patients with severe kidney problems, including those who are undergoing renal replacement therapy. The approval of plazomicin provides clinicians with an additional option for treating adults with cUTIs, particularly those caused by multidrug-resistant Gram-negative rods [40].

The 16S rRNA ribosomal methyltransferase activity prevents the binding of almost all aminoglycoside molecules, including plazomicin, causing high resistance levels. Previous studies have shown that these methyltransferases are uncommon in U.S. hospitals, with only 6 (0.07%) observed in a 2-year surveillance of 8,000 isolates. The absence of 16S rRNA ribosomal methyltransferases led plazomicin to exhibit *in vitro* action versus *Enterobacterales* isolates that do not transport these methyltransferases. A comparison of the activity against CRE, MDR, and XDR isolates shows that plazomicin is significantly more effective than gentamicin and tobramycin. Amikacin, gentamicin, and tobramycin activities were much lower against CRE, MDR, and XDR isolates when using PK/PD parameters similar to those used to determine plazomicin breakpoints [32, 35, 41].

Carbapenem-resistant Enterobacteriaceae (CRE)

Enterobacteriaceae, a rod-shaped bacterium belonging to the Gamma-proteobacteria class in the *Pseudomonadota* phylum, comprises 68 different types and 355 species, which include the well-known *Escherichia*, *Klebsiella*, and *Enterobacter* [43]. Some Enterobacteriaceae bacteria have resisted carbapenem, an antibiotic typically used as a last resort for treating bacterial infections. This resistance has led to the creation of Carbapenem-resistant Enterobacteriaceae (CRE), a group of bacteria considered a global priority pathogen due to its resistance to multiple drugs and ability to spread resistance genes through mobile genetic elements. CRE can cause a broad spectrum of infections, including urinary tract infections, bloodstream infections, and pneumonia. Nevertheless, their resistance to multiple antibiotics makes treating these infections challenging and can result in complications and an increased risk of death [42, 44].

The spread of CRE is a significant public health concern, and healthcare facilities are mainly at risk due to the high use of antibiotics, invasive procedures, and close patient contact [45].

In 2020, the scientific community modified taxonomy by introducing a new scientific order called "Enterobacterales." Currently, "Enterobacteriaceae" is a family that belongs to the Enterobacterales order, along with other families such as Morganellaceae, Budvicaceae, Pectobacteriaceae, Yersiniaceae, Hafniaceae, Morganellaceae, and Erwinaceae [46].

The Enterobacteriaceae can survive in hospitals and cause hospital-acquired infections (HAIs), which can lead to increased morbidity and mortality rates. *Klebsiella pneumoniae, Escherichia coli*, Proteus, and *Enterobacter* spp. are more frequently isolated from clinical specimens than other Enterobacteriaceae spp—species; in hospital precincts, *E. coli, Klebsiella* spp., and *Enterobacter* cause respiratory tract, blood, skin diseases, and urinary tract infections (UTIs). The presence of antibiotic resistance sequences, especially carbapenemase-producing genes, is the main reason for the antibiotic resistance in Enterobacteriaceae, which entails extended hospitalization and limitations in antibiotic treatment such as broad-spectrum third-generation penicillin, cephalosporins, fluoroquinolones, and carbapenems. It has been reported that the dissemination from South America to other parts of the world of the most described carbapenemases in *Enterobacterales* was through clones, plasmids, and transposons [47].

The Centers for Disease Control and Prevention (CDC) have reported a marked increase in Carbapenem-Resistant Enterobacteriaceae (CRE) infections in the United States. In 2017, a total of 13,100 cases of CRE were recorded in hospitalized patients, resulting in 1,100 deaths. Patients requiring medical devices such as catheters and undergoing long-term antibiotic treatment are at a higher risk of contracting CRE infections. Furthermore, colonization is an additional risk factor for diseases caused by CRE. The bacteria can be present in asymptomatic individuals, who can transmit the bacteria to other patients, thus exacerbating the problem. Therefore, it is essential to prevent CRE colonization to avoid the spread associated with these infections [48].

As CRE becomes more common globally over time, it is crucial to distinguish between carbapenemase-producing CRE (CP-CRE) and non-carbapenemas--producing CRE (nonCP-CRE). A strain only needs to acquire one carbapenemase to become CP-CRE, but non-CP-CRE can have various resistance mechanisms that are often difficult to identify.

According to the Antibiotic Resistance Laboratory Network (ARLN), in the USA, 68% of CRE isolates cataloged in 2017 were non-CP-CRE, which is a worrying number. *Klebsiella aerogenes, K. pneumoniae, Escherichia coli*, E. cloacae species complex, and *S. marcescens* [49] are the most prevalent carbapenem-resistant *Enterobacteriaceae* species.

A combination of antibiotics is recommended to combat antibiotic-resistant CRE strains rather than relying on monotherapies. Studies have shown that combinations such as polymyxin B and rifampicin or polymyxin B and imipenem can effectively eliminate these strains [50]. A recent study found that a triple-drug combination of polymyxin B, rifampin, and meropenem at 1/4 of their MICs was highly effective in treating *E. coli* and *K. pneumoniae* clinical isolates that produced KPC-type enzymes [51].

Interactions of colistin (polymyxin E) and imipenem were also examined in time-kill experiments against 42 VIM-producing *K. pneumoniae* (Verona integron-encoded metallo-β-lactamase, VIM) hospital isolates. This combination generally exhibited improved bactericidal activity against isolates susceptible to either agent or colistin alone. Similarly, the combination of colistin and tigecycline was synergistic against 50% of colistin-susceptible isolates, while it was indifferent against the remaining 50%. However, the colistin and tigecycline combination showed a 55.6% antagonistic effect for isolates non-susceptible to colistin. The study also found that polymyxin B plus rifampin and polymyxin B plus doxycycline showed pronounced synergy, while less pronounced synergy was observed with polymyxin B and tigecycline. However, there was no synergy between polymyxin B and other antimicrobial agents, including imipenem and gentamicin. The study concluded that the outbreak was characterized by multiple clones, with one dominant strain affecting the majority of hospitals. Therefore, different clones responded to antibiotic combinations to varying extents. The study suggests that polymyxin B–rifampicin or tigecycline alone may help treat the infection. Collaborative infection control actions will be essential to contain these pathogens' spread.

According to a study, combining fosfomycin with meropenem showed synergy in 64.7% of KPC-producing *K. pneumoniae* isolates, while fosfomycin and colistin combination showed only 11.8% in the identical isolates. However, combining gentamicin with fosfomycin did not show a synergic effect. According to a similar study, fosfomycin with meropenem, colistin, and gentamicin combinations avoided the resistance to fosfomycin in 69.2%, 53.8%, and 81.8% of analyzed isolates, respectively [52]. Additional studies also demonstrated comparable results, displaying the interaction of fosfomycin, including imipenem, meropenem, doripenem, colistin, netilmicin, and tigecycline in 74%, 70%, 74%, 36%, 42%, and 30% of producing *K. pneumoniae* isolates, respectively [52].

Time-kill assays were used to compare the effectiveness of aztreonam and carbapenems. Aztreonam is not broken down by metallo-β-lactamase (MBLs), making it a potentially helpful agent against MBL producers. A time-kill study has compared the *in vitro* activity of aztreonam and carbapenems versus VIM--producing ESBL-negative *K. pneumoniae* isolates. The outcomes showed that aztreonam displayed 24 hours of slow bactericidal action. Although carbapenems cause faster bacterial killing during the first 6 hours, regrowth occurred, reaching antibiotic-free controls at 24 hours, particularly in cases where KPCs predominate (Figs. **5** and **6**) [53].

In a recent study, all 32 strains of *Klebsiella* pneumoniae producing carbapenemases showed resistance to several antibiotics, such as carbapenems, ceftazidime, and others. Interestingly, just 9.3% of the strains were resistant to fosfomycin. However, when the sensitivity trial was replicated, deprived of G6-P, 82.7% of the fosfomycin-susceptible strains were found to be resistant to fosfomycin. According to the study, 46.8% of the identified carbapenemases were NDM-1, 18.7% were OXA-48, 3.1% were KPC, and 21.8% were a combination of NDM-1 and OXA-48. It is worth noting that only one of the resistant strains had OXA-48, and no viable genes were found in the two resistant Strains. The study suggests that IV fosfomycin could be an effective alternative treatment for infections caused by common resistant strains. The effectiveness of IV fosfomycin was assessed using the agar dilution method, which was repeated without G6-P to determine its impact on sensitivity results [54].

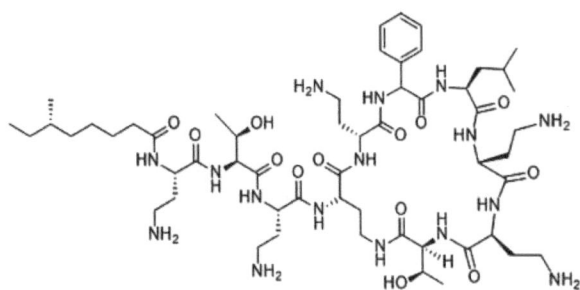

Polymyxin B1

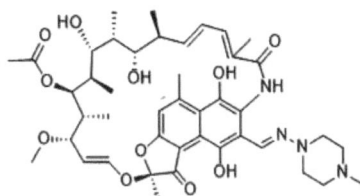

Rifampicin

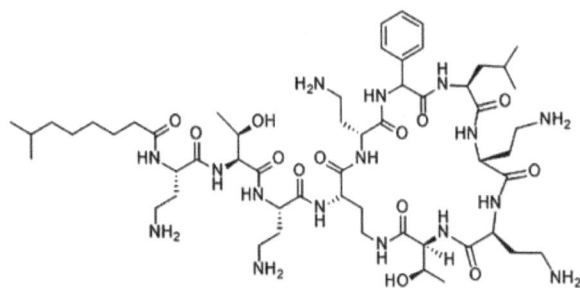

Polymyxin B2

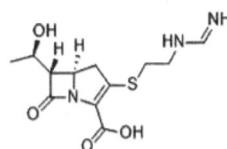

Imipenem

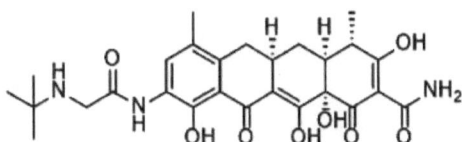

Tigecycline

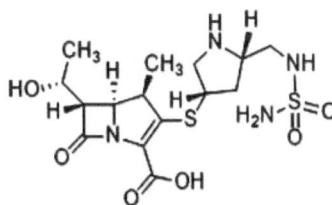

Doripenem

Fig. (5). Chemical Structures of Polymyxin B, Rifampicin, Imipenem, Tigecycline, Doripenem, Doxycycline, Colistin, and Gentamicin.

Fig. (6). Chemical structures of Fosfomicyn, Meropenem, Netilmicin, and aztreonam.

Urinary tract infections (UTIs) are more frequent among women and are more likely to occur as women age. However, due to increased antibiotic resistance and the limited availability of new agents, UTI treatment has become more challenging. A cross-sectional study was conducted in Belgium, Italy, Spain, Russia, and the UK to collect urinary isolates from non-hospitalized women at 20 locations. Relevant antibiotics were used to perform disk diffusion experiments on the bacteria. For quality control, a central laboratory retested isolates found to be resistant to fosfomycin and every tenth isolate. All non-Russian sites were included in the quality control test.

The study analyzed 2,848 isolates and found that *Escherichia coli* (2064; 72.5%) was the most common bacterium detected, while *Klebsiella* spp (275; 9.7%) and *Proteus* spp. (103; 3.6%) were less frequently found. Nitrofurantoin (98.5%), fosfomycin (96.4%), and mecillinam (91.8%) showed activity against >90% of *E. coli* isolates. Furthermore, fosfomycin and nitrofurantoin were effective against >90% of cephalosporin-resistant *E. coli*. Of the 143 *E. coli* documented locally by disk tests, 138 (96.5%) were susceptible to minimum inhibitory concentration (MIC) tests. However, only 29 out of the 58 isolates reported as resistant by local disk tests were confirmed to be resistant by MIC tests. The study found that *Escherichia coli* was highly susceptible to fosfomycin, nitrofurantoin, and mecillinam, all of which have been used for over 30 years. Therefore, for uncomplicated UTIs in women, recommendations supporting fosfomycin remain microbiologically valid [55].

Candida auris

The availability of limited antifungals and the development of multi-drug resistance in fungal pathogens have become severe concerns in the health sector. Although several cellular, molecular, and genetic mechanisms have been proposed to explain the drug resistance mechanism in fungi, a complete understanding of this mechanism still needs to be improved.

Several species of *Candida* are a group of fungi that are resistant to antifungal medications and can cause infections ranging from mild oral and vaginal candidiasis to severe invasive infections. It is a common problem for hospitalized patients, with about 7% of bloodstream infections resistant to antifungals. The main classes of antifungal medications used to treat *Candida* infections are azoles and echinocandins.

Recent research has revealed that *Candida* is a type of yeast that looks similar to other species but is genetically different. For instance, *C. auris* is haploid and differs from other pathogenic diploid *Candida* species, such as *Candida albicans* and *Candida glabrata*. This genetic difference allows *C. auris* to develop drug resistance, which is rare among other *Candida* species. As a result, medical mycology has turned its attention towards *C. auris* due to its ability to resist [56].

Candida auris was first identified in 1996 and has since been recognized as a global nosocomial pathogen. This yeast is challenging to treat and has high clonal transmission within and between hospitals. It rapidly affects patients and persistently colonizes the skin, but the exact transmission mode is still unclear. *C. auris* is also multidrug-resistant, making it challenging to manage. Unfortunately, there is currently no recommended approach for decolonizing this yeast [57].

Amphotericin B (Fig. **7**) is a potent medication for treating severe fungal infections. It belongs to the class of polyene macrolide antibiotics and is administered intravenously. It forms pores in the cell membrane by binding to ergosterol, inducing the death of the fungal cells. Amphotericin B effectively treats fungal infections, such as aspergillosis, blastomycosis, candidiasis, coccidioidomycosis, and cryptococcosis.

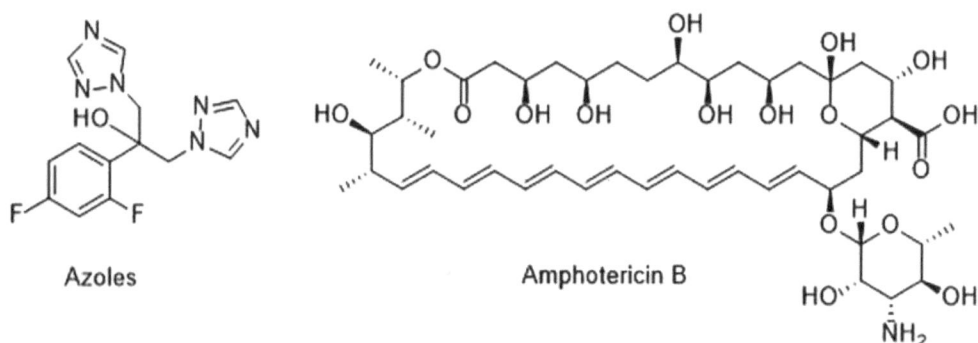

Fig. (7). Chemical structures of Azoles and Amphotericin B.

Despite its effectiveness in treating severe, life-threatening fungal infections, amphotericin B should only be prescribed as a last resort due to its significant side effects, including fever, chills, headache, nausea, vomiting, and kidney damage. However, recent reports have shown that certain strains of the fungus *C. auris* are resistant to this drug. Mutations in the ERG6 gene of *C. auris* were found to be the first known case of clinical amphotericin B resistance in this fungus. This new information is crucial in understanding antifungal resistance and its implications for public health [58].

A recent publication focused on discovering new molecules to combat pathogenic *Candida* species. The study explored the fungicide effects of stilbene-type derivatives against drug-resistant strains of *C. albicans* and *C. auris*. These new molecules were derived from 4.4' dihydroxy azobenzene, structurally like antifungal stilbene derivatives in *Agaricus xanthodermus* (Yellow stainer). The study showed that these compounds effectively prevented the growth of both fluconazole-susceptible and fluconazole-resistant strains of *Candida albicans* and *Candida auris* [59].

One method to treat *C. auris* involves synthesizing and testing an anti-Hyr1p MAb monoclonal antibody explicitly targeting drug-resistant *C. auris* strains. The results of this test showed that the anti-Hyr1p MAb prevented the formation of biofilms and also helped macrophages eliminate *C. auris* more effectively. When tested *in vivo*, the anti-Hyr1p MAb successfully protected 55% of mice from lethal systemic *C. auris* infection and significantly reduced the fungal burden [60].

Recent evidence suggests that lemongrass oil can eliminate the *C. auris* fungus. The terpenoids present in lemongrass oil, specifically the citral isomers, have been shown to decrease membrane fluidity, which makes *Candida* spp and drug-resistant bacteria more vulnerable. This finding implies that lemongrass oil's fungicidal mechanism is directly related to the decrease in membrane fluidity. In addition, a study has shown that lemongrass oil can be effectively used with fluconazole, amphotericin B, flucytosine, and micafungin to enhance their activities (Fig. **8**). These findings suggest lemongrass oil can disinfect communal spaces, promoting optimal health [61].

Isavuconazole: This second-generation triazole derivative with broad-spectrum activity versus yeasts, dimorphic fungi, and molds is distinguished by predictable pharmacokinetics and a good safety profile. Isavuconazole is generally well-tolerated by patients and has fewer drug interactions. It works by inhibiting cytochrome P450-dependent lanosterol 14-α-demethylase, which is necessary to produce ergosterol, a crucial component of the fungal membrane [62]. Isavuconazole is as effective as voriconazole in treating invasive aspergillosis, and it can be used as an alternative therapy for mucormycosis. It is also suitable for step-down therapy in cases of invasive candidiasis. However, cross-resistance with other triazoles is common [63]. Invasive fungal diseases can be severe infections caused by fungal pathogens affecting different body parts, including the lungs, blood, brain, and other organs. Opportunistic fungi can infect individuals with weakened immune systems, such as HIV/AIDS, cancer, or transplant recipients receiving immunosuppressive therapies. Isavuconazole (ISA) is an effective invasive fungal disease (IFD). Invasive fungal infections, such as invasive aspergillosis (IA), invasive mucormycosis (IM), and intrusive candidiasis (IC), are associated with high morbidity and mortality rates, particularly in immunocompromised patients. ISA has emerged as a highly effective treatment option for these infections. Clinical trials have demonstrated that ISA is as effective as voriconazole and liposomal amphotericin B (L-AmB) in treating IA and IM and is recommended as a first-line treatment. Additionally, ISA can be used as an alternative to de-escalation therapy after initial drug treatment for IC.

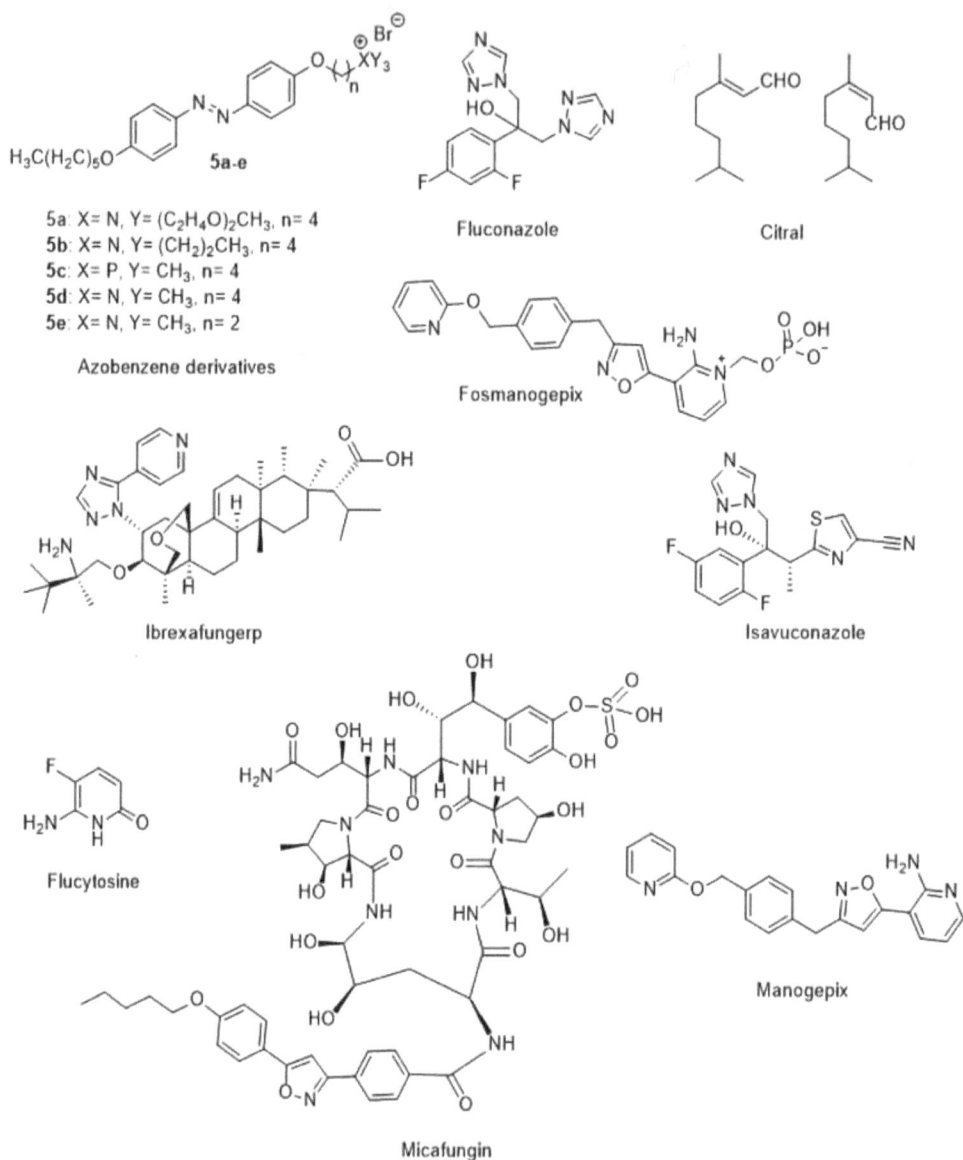

5a: X= N, Y= (C₂H₄O)₂CH₃, n= 4
5b: X= N, Y= (CH₂)₂CH₃, n= 4
5c: X= P, Y= CH₃, n= 4
5d: X= N, Y= CH₃, n= 4
5e: X= N, Y= CH₃, n= 2

Azobenzene derivatives

Fluconazole

Citral

Fosmanogepix

Ibrexafungerp

Isavuconazole

Flucytosine

Manogepix

Micafungin

Fig. (8). Chemical structures of Azobenzene derivatives, Fluconazole, Citral, Ibrexafungerp, Fosmanogepix, Isavuconazole, Flucytosine, Micafungin, Manogepix.

While further research is necessary to determine ISA's effectiveness in preventing fungal infections in high-risk patients, it has been proven safe and effective for invasive fungal disease (IFD) prevention [64]. These findings suggest that ISA may be a promising therapeutic option for patients at risk of IFD.

Fosmanogepix is a fosfo-derivative of manogepix that acts as a prodrug and inhibits the fungoid enzyme Gwt1. This enzyme is a potential target for antifungal drugs, as inhibiting Gwt1 can disrupt fungal cell membrane integrity and lead to cell death. Inhibition of Gwt1 has been shown to have antifungal activity against several pathogenic fungal species, including *C. albicans* and *Aspergillus fumigatus*. Fosmanogepix is a pharmacological agent that operates through a unique mechanism of action, rendering it efficacious against a wide variety of drug-resistant fungal strains, including *Candida,* resistant to echinocandin, and *Aspergillus*, resistant to azole. It also exhibits activity against *Scedosporium lomentospora* prolific and *Fusarium*, pathogens that inherently resist other drug groups. Fosmanogepix has revealed a remarkable ability to mitigate disseminated diseases caused by *Candida* bacteria, *Coccidioides immitis*, and *Fusarium solani*. It improved the condition of the lungs, which were affected by *Aspergillus fumigatus, Aspergillus flavus, Staphylococcus* bacteria, and *Rhizopus arrhizus* in animal models. Clinical studies have confirmed that fosmanogepix has high bioavailability—over 90%. This enables easy switching between oral and intravenous forms without affecting blood levels. Fosmanogepix also has a favorable drug interaction profile and tolerability. Extensive tissue distribution renders it an attractive option for managing invasive fungal infections [65].

Ibrexafungerp is a new semisynthetic fungicidal derivative of the hemiacetal triterpene glycoside enfumafungin. It acts as a β-D-glucan synthase inhibitor with similar but not identical binding sites to echinocandins in the enzyme's catalytic regions Fks1p and Fks2p [66]. The susceptibility of echinocandin-resistant *Candida* strains to ibrexafungerp is related to hot spot mutations in Fksp in Candida [67]. On the other hand, analyzing *C. glabrata* strains with resistance to echinocandins and susceptibility to ibrexafungerp showed that ibrexafungerp only has a partial overlap in the Fksp binding sites of echinocandins in the enzyme β-D-glucan synthase [66]. Ibrexafungerp showed good tissue penetration and high protein binding, although it has little penetration into the CNS, like echinocandins [68]. It is designated for caring for postmenarchal pediatric females with vulvovaginal candidiasis (VVC) [69].

Ibrezanfungerp intravenous formulation is currently in late-stage clinical trials for the treatment of fungal infections. It has shown antifungal activity against *C. auris* and *Aspergillus* species resistant to azoles and echinocandins—strains, both *in vitro* and *in vivo* [70]. Ibrexafungerp has demonstrated promising *in vitro* activity against fungal pathogens such as *Candida* spp., including multidrug-resistant *C. glabrata*, biofilm producer strains, and *C. auris* [71, 72].

Previous reports show that some *Candida auris* strains are drug-resistant to some fungicidal compounds. To inhibit *C. auris* infection, isavuconazole- echinocandin combinations were tested. All assortments of isavuconazole with echinocandins resulted in a general synergistic interaction. The excellent distribution of these combinations generates an attractive option for managing invasive fungal infections. Synergism and fungistatic action were still accomplished with combinations that involved low concentrations of isavuconazole (≥ 0.125 mg/L) and (≥ 1 mg/L) echinocandin. Time-kill curves exposed that after synergy was accomplished, the combinations of higher concentrations of the drugs did not improve antifungal activity. This work shows promising results regarding combining isavuconazole with echinocandins for treating *C. auris* infections [73, 74].

It is important to note that treating *Candida auris* infections can be challenging, and a combination of antifungal drugs may be necessary to achieve a successful outcome. The treatment plan should be individualized based on the specific characteristics of the infection and the patient's medical history (Table **2**).

Table 2. Drugs against *Candida auris*.

Compound	Observations	Reference
Amphotericin B	It works by binding to the fungal cell membrane, disrupting its structure and function and ultimately leading to the death of the fungus. However, due to its high toxicity, it is only used in aggressive *Candida* infections.	[56]
Azobenzene derivatives	These compounds prevented the growth of both fluconazole□susceptible and fluconazole□resistant *Candida albicans* and *C. auris* strains. Further, *in vivo* studies are required to confirm the potential therapeutic value of these compounds	[57]
Lemon grass oil	Citral has been linked to increased susceptibility to drug-resistant bacteria and Candida by decreasing membrane fluidity.	[59]
Isavuconazole	In 2020, the U.S. FDA approved an antifungal drug for treating *Candida auris* infections.	[61]
Fosmanogepix	An investigational antifungal medicine has displayed potential in early clinical trials to treat *Candida auris* infections. This medicine targets a pathway different from other antifungal drugs, which may make it effective against fungus strains resistant to other treatments.	[63]
Ibrexafungerp	It is currently being studied in clinical trials to treat fungal infections, including those caused by *Candida auris*.	[64]

Clostridioides difficile (C. difficile)

Clostridiodes difficile, a microorganism discovered in 1935, has been linked to antibiotic-associated diarrhea since 1978. Its increasing resistance to antimicrobial treatments poses a significant threat to global healthcare systems [75]. In addition to causing health problems, *C. difficile* is a reservoir for antimicrobial resistance genes that could spread to other pathogens. There are some proposed mechanisms of *C. difficile* resistance; evidence shows that previously unknown mechanisms, such as plasmid-mediated resistance, may play a crucial role in its antimicrobial resistance. In 2017, *C. difficile* infection caused around 223,900 hospitalizations and 12,800 fatalities in the US alone, making it a significant public health concern globally. Broad-spectrum antibiotics like fluoroquinolones, cephalosporins, and clindamycin (Fig. **9**) disrupt the natural intestinal microbiota, leading to accelerated colonization of the gastrointestinal tract by *C. difficile* and the development of *C. difficile* infections (CDI).

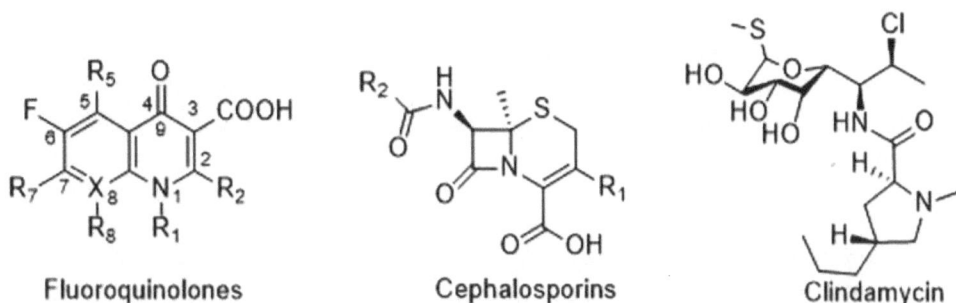

Fluoroquinolones   Cephalosporins   Clindamycin

Fig. (9). Chemical structures of Fluoroquinolones, Cephalosporins, and Clindamycin.

Currently, only a limited number of antibiotics are effective in treating CDI. Vancomycin and fidaxomicin are the recommended first options to treat an initial episode of CDI and for recurrences. Although it is no longer the first option, metronidazole can treat non-severe CDI in adults and children. Recently, rifamycins like rifaximin were explored as an additional therapy for treating CDI (Fig. **10**). Unfortunately, resistance or decreased susceptibility to these antibiotics has been reported, making it a significant challenge for patients and clinicians due to the lack of available treatment options for CDI [76].

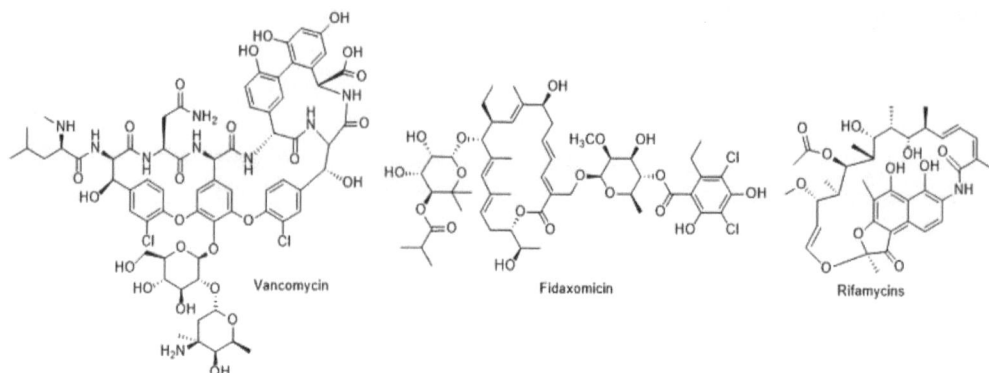

Fig. (10). Chemical structures of Vancomycin, Fidaxomicin and Rifamycin.

Recent research has confirmed that *C. difficile* toxins A and B are the primary cause of colitis linked to antibiotic use. Unfortunately, almost all antibiotics have been found to trigger this type of colitis [77].

While CDI has historically been associated with changes in the normal microbiota of the intestines following antibiotic use, it has become increasingly prevalent over the past ten years in low-risk individuals who have not recently taken antibiotics. *Clostridium difficile* is now the eighth most reported microorganism in healthcare-related infections, and its incidence is on the rise in most countries.

Clostridiodes difficile ribotype 27 strain is more dangerous due to the deletion of 18 bases in the TCDC gene, which regulates the pathogenicity locus and the presence of the binary toxin. This strain is also resistant to quinolones and has caused a threefold increase in hospitalizations compared to previous years. Additionally, older adults over 65 are more susceptible to lethal CDI. A report from Finland found that the 30-day fatality rate for community-acquired CDI was 3.2%, while hospital-acquired CDI had a fatality rate of 13.3% [78].

The treatment against *C. difficile* is complicated when there is a recurrence of infection (rCDI). So far, the best therapy against rCDI is fecal microbiota transplantation (FMT). However, FMT is unavailable everywhere, and patients must be eligible for the procedure [79].

On the other hand, bezlotoxumab is a fully humanized antibody against *C. difficile* toxin B, and it is indicated for preventing rCDI in at-risk patients [80]. Nevertheless, other treatment options are still needed.

There are several limitations to treating *C. difficile* infections (CDI). Oral metronidazole is recommended for mild to moderate infections; however, in two of three clinical trials, it was less active than oral vancomycin for complicated CDI. Oral vancomycin disrupts the normal intestinal flora and has a four-time--daily dosing regimen.

According to the latest report of the European Society that Study Infectious Diseases (ESCMID), metronidazole should only be used if no fidaxomicin or vancomycin is available. If feasible, fidaxomicin is the preferred treatment option for the initial and first recurrence of CDI. FMT or bezlotoxumab with standard care antibiotics (SoC) are the preferred options for CDI's second or further recurrence. Bezlotoxumab, in combination with SoC, is recommended as the first resort for CDI if fidaxomicin is used to manage the initial episode. When fidaxomicin is not feasible, bezlotoxumab is considered an additional treatment option to vancomycin for a CDI episode with a high risk of recurrence. It is important to note that the treatment strategy for an individual patient should be determined based on the risk for recurrence rather than the severity of the disease [81].

Although oral vancomycin and metronidazole have been recommended against *C. difficile*-induced infections, these drugs are associated with vancomycin-resistant *enterococci* (VRE) colonization. Approximately 20% to 30% of patients experience recurrent disease, with higher rates observed in those with multiple episodes and high-risk subgroups (oncology, renal failure, concomitant antibiotics, advanced age, previous episodes of CDI).

Fidaxomicin is a newer macrocyclic lactone antibiotic specifically designed to target *C. difficile*. It inhibits RNA polymerase, preventing transcription. Fidaxomicin (and its active metabolite OP-1118) exhibits a prolonged post-antibiotic effect. It has been reported that fidaxomicin showed a clinical response equal to that achieved by oral vancomycin at the end of CDI treatment. Fidaxomicin is as safe as oral vancomycin but superior in achieving sustained clinical response for CDI in patients infected with strains other than BI/NAP1/027. Due to the limited data availability, it is essential to exercise caution when employing fidaxomicin monotherapy to treat severe, complicated CDI. The cost-effectiveness of fidaxomicin administration, as opposed to oral vancomycin, hinges on the acceptable willingness-to-pay threshold per quality-adjusted life year (QALY) to evaluate its cost-effectiveness. This is due to the significantly higher procurement cost of fidaxomicin, which may affect the economic feasibility of its usage [82].

Neisseria gonorrhoeae (N. gonorrhoeae).

Statistics about Sexually transmitted infections (STIs) showed that more than 1 million STIs are acquired daily. WHO, in 2020, estimated 374 million new infections, of which approximately 25% were STIs: *Chlamydia,* 129 million; *gonorrhea,* 82 million; *syphilis,* 7.1 million; and *trichomoniasis,* 156 million. They are estimating that in 2020, 82.4 million people were newly infected with gonorrhea caused by *Neisseria gonorrhoeae.* Although gonorrhea infections continue to be curable when treated with antibiotics, there are some concerns since gonorrhea antimicrobial resistance (AMR) has increased over the past 50 years. Gonorrhea has developed resistance to many classes of antibiotics, including quinolones and early-generation cephalosporins. It has even resisted some of the last line of defense antibiotics, making it a multidrug-resistant pathogen [83].

In 2021, the recent outcomes from WHO's global antimicrobial resistance surveillance (GASP/GLASS) for *Neisseria gonorrhoeae* isolated from 2017-2018 were available. The report corroborated that resistant gonococcal strains are widespread globally, as data were collected from 73 countries. The sexually transmitted bacteria *gonorrhea* is the second most reported infection in the European Union/European Economic Area (EU/EEA). Numerous cases were reported in 2019, yielding an increase of 17% in one year.

The current recommended treatment for gonorrhea involves a combination of two antibiotics: ceftriaxone (an injectable cephalosporin) and azithromycin (a macrolide). The present approach incorporates two antibiotics to augment therapeutic efficacy while reducing the risk of developing antibiotic resistance.

Ceftriaxone is a third-generation cephalosporin antibiotic administered by injection. It inhibits bacterial cell wall synthesis, ultimately leading to the bacteria's death. Ceftriaxone is highly effective against *Neisseria gonorrhoeae.*

The current recommended treatment for uncomplicated gonorrhea is only one dose of 500 mg of intramuscular ceftriaxone plus a single dose of 1g of oral azithromycin. This treatment regimen is highly effective and recommended by several international Health Organizations. However, antibiotic resistance is a growing concern in treating gonorrhea, and the currently recommended treatment regimen may not be effective against all strains of the bacteria.

Solithromycin is a macrolide antibiotic investigated for the treatment of gonorrhea. It has shown promising activity against drug-resistant strains of *N. gonorrhoeae in vitro* and animal studies. Clinical trials are ongoing to evaluate its safety and efficacy in human clinical trials.

Zoliflodacin belongs to a new class of antibacterial agents called spiro pyrimidinetriones. The FDA has granted it fast-track status, and it has become the first new antibiotic approved for treating gonorrhea in over a decade. This compound works by inhibiting bacterial DNA synthesis. It is effective against both drug-susceptible and drug-resistant strains of *N. gonorrhoeae*. Zoliflodacin is a highly effective inhibitor of bacterial class II topoisomerases with critical locations in bacterial gyrase. Its mechanism of action differs from that of fluoroquinolones, and it has been found to have a lower incidence of antagonism and vigorous antibacterial activity against *N. gonorrhoeae*, including multidrug-resistant strains. The minimum inhibitory concentrations (MICs) of zoliflodacin against *N. gonorrhoeae* have been found to range from ≤0.002 to 0.25 µg/mL [84].

Lefamulin is an antibiotic that belongs to the pleuromutilin class and is considered a state-of-the-art antibiotic. It is prescribed to treat adult bacterial infections, specifically pneumonia acquired in the community (NAC) caused by susceptible bacteria. The Food and Drug Administration of the United States (FDA) approved this medication in August 2019 for adult patients. Lefamulin inhibits the synthesis of bacterial proteins. It does this by joining the 50s subunit of the bacterial ribosome, interrupting protein synthesis and ultimately leading to the death of bacteria. Lefamulin's advantage is that it is active against many bacteria, including strains resistant to common antibiotics such as penicillin and fluoroquinolone. In addition, it can be administered both orally and intravenously, making it a convenient option for treating NAC in hospitalized and outpatient patients. It has been shown that lefamulin is effective against some strains of *Neisseria gonorrhoeae*. However, more studies are still needed to determine its effectiveness and safety in treating gonorrhea. It has evaluated the activity of lefamulin, inhibiting the effect of efflux pump inactivation besides protein synthesis and growth on clinical gonococcal isolates and connection strains that contain considerable multidrug-resistant and drug-resistant isolates. Lefamulin (Fig. **11**) showed decisive activity vs. all gonococcal isolates, and no marked cross-resistance to other antimicrobials was determined. Additional analyses of lefamulin are demanded, including *in vitro* selection and opposition mechanisms, pharmacokinetics/pharmacodynamics, optimal dosing, and performance in randomized controlled trials [85].

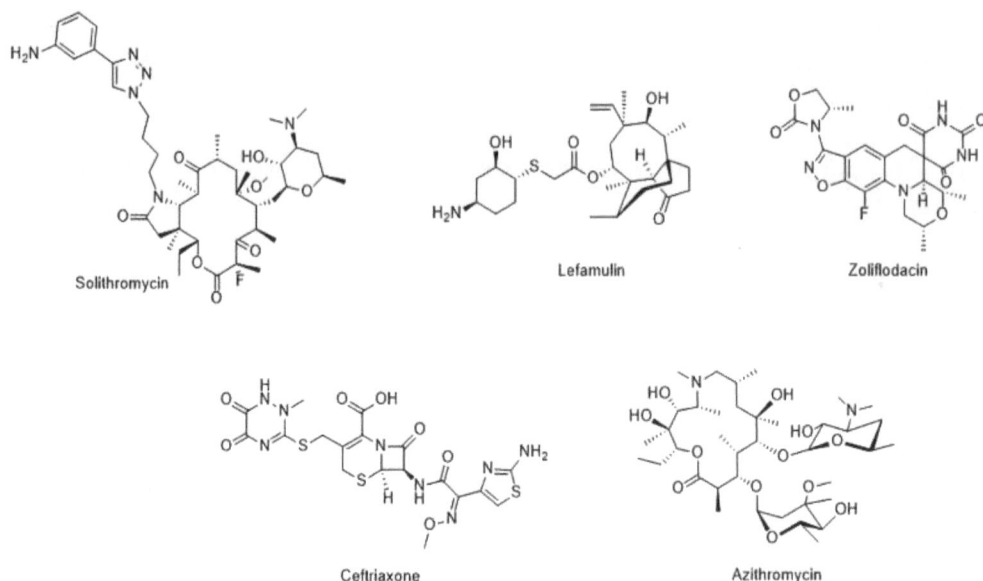

Fig. (11). Chemical structures of Solithromycin, Lefamulin, Zoliflodacin, Ceftriazone and Azithromycin.

It is important to note that developing new drugs is only part of the solution to the problem of drug-resistant *N. gonorrhea*. Prevention measures, such as safe sex practices, are crucial to controlling the spread of the infection. Additionally, public health efforts to monitor and track the emergence of drug-resistant strains of N. gonorrhea are essential to guiding treatment strategies and preventing the further spread of the infection.

Serious Threats

• *Multidrug-resistant Pseudomonas aeruginosa* (*P. aeruginosa*)

Pseudomonas aeruginosa, a Gram-negative bacterium, thrives in damp environments, soil, and water. It threatens individuals with weakened immune systems, burns or injuries, and cystic fibrosis. Additionally, *P. aeruginosa* is an essential contributor to bioremediation, as it can break down a wide range of organic compounds, making it an effective tool in cleaning contaminated soil and water [86]. However, managing infections caused by *P. aeruginosa* can be challenging due to its antibiotic resistance. Furthermore, this bacterium can form biofilms, protecting it from the immune system and antibiotics. The world is seeing an increase in antibiotic resistance rates in *P. aeruginosa*. This species can develop multidrug resistance (MDR) through various mechanisms, including overexpression of specific efflux pumps like MexAB-OprM, MexCD-Oprj, and

MexEF-OprN. These pumps can inhibit the effectiveness of macrolides, aminoglycosides, sulfonamides, fluoroquinolones, tetracyclines, or β-lactams. The deletion of porins (OprD) can inhibit imipenem and meropenem, while β-lactamases like PSE-1, PSE-4, AmpC, and Metallo-β-lactamases can inhibit penicillins, third-generation cephalosporins, piperacillin, and carbapenems. Additionally, genes like *rmtA*, *rmtB*, and *armA*, which code for 16S rRNA methylases, can inhibit aminoglycosides, and target mutations like the quinolone resistance-determining region can inhibit fluoroquinolones [87].

There is no globally accepted definition of multidrug resistance, making it hard to compare different studies. For example, using the wrong treatment method is linked to higher mortality rates in cases of *P. aeruginosa* infections. Waiting too long to administer the proper treatment can also prolong hospitalization and worsen the disease. Moreover, limited options for effective antimicrobial therapy may lead to poor clinical outcomes in cases of MDR infections [87].

There is a debate about whether MDR *P. aeruginosa* infection significantly impacts patient outcomes. While it was previously believed that MDR pathogens are less harmful, recent studies challenge this idea. The association between resistance mechanisms and virulence factors can be better understood through molecular investigations and animal studies. In clinical scenarios where ethical concerns arise, animal studies can help fill knowledge gaps. Currently, the most effective treatment for MDR *P. aeruginosa* is a combination of colistin and another antibiotic. However, there are still several classes of antibiotics (Fig. **12**) that can be effective against *P. aeruginosa*, including:

- Aminoglycosides, such as gentamicin and tobramycin, bind to the bacterial ribosome, interfering with protein synthesis.
- Carbapenems, broad-spectrum antibiotics such as imipenem and meropenem, penetrate the bacterial cell wall and inhibit the synthesis of the cell wall.
- Cephalosporins, such as ceftazidime and cefepime, are broad-spectrum antibiotics inhibiting cell wall synthesis and protein synthesis.
- Fluoroquinolones, such as ciprofloxacin and levofloxacin, inhibit DNA synthesis.

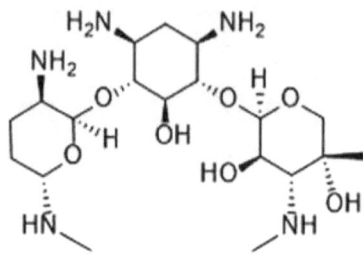

Gentamicin

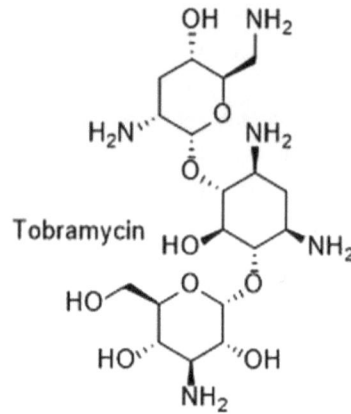

Tobramycin

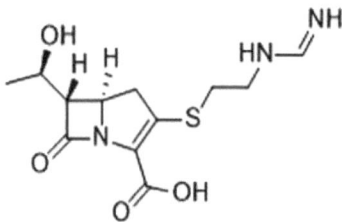

Imipenem

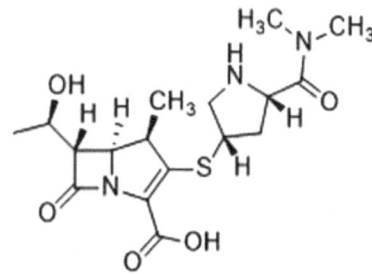

Meropenem

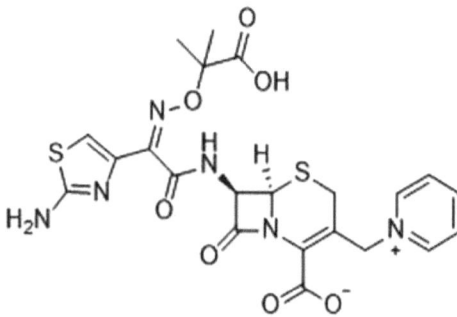

Ceftazidime

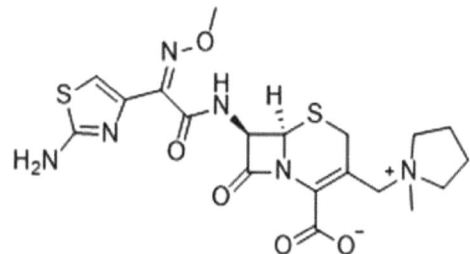

Cefepime

Fig. (12). Chemical structures of Gentamicin, Tobramycin, Imipenem, Meropenem, Ceftazidime, Cefepime, Ciprofloxacin, and Levofloxacin.

Antibiotic-resistant strains of *P. aeruginosa* are becoming increasingly prevalent and pose a significant challenge in terms of treatment. *P. aeruginosa* can become antibiotic-resistant by producing efflux pumps that can expel antibiotics from the bacterial cell or by synthesizing enzymes that can break down antibiotics. However, despite these challenges, several drugs are still available to treat infections caused by antibiotic-resistant *P. aeruginosa*. Fig. (**13**) highlights some of the most used medications for this purpose.

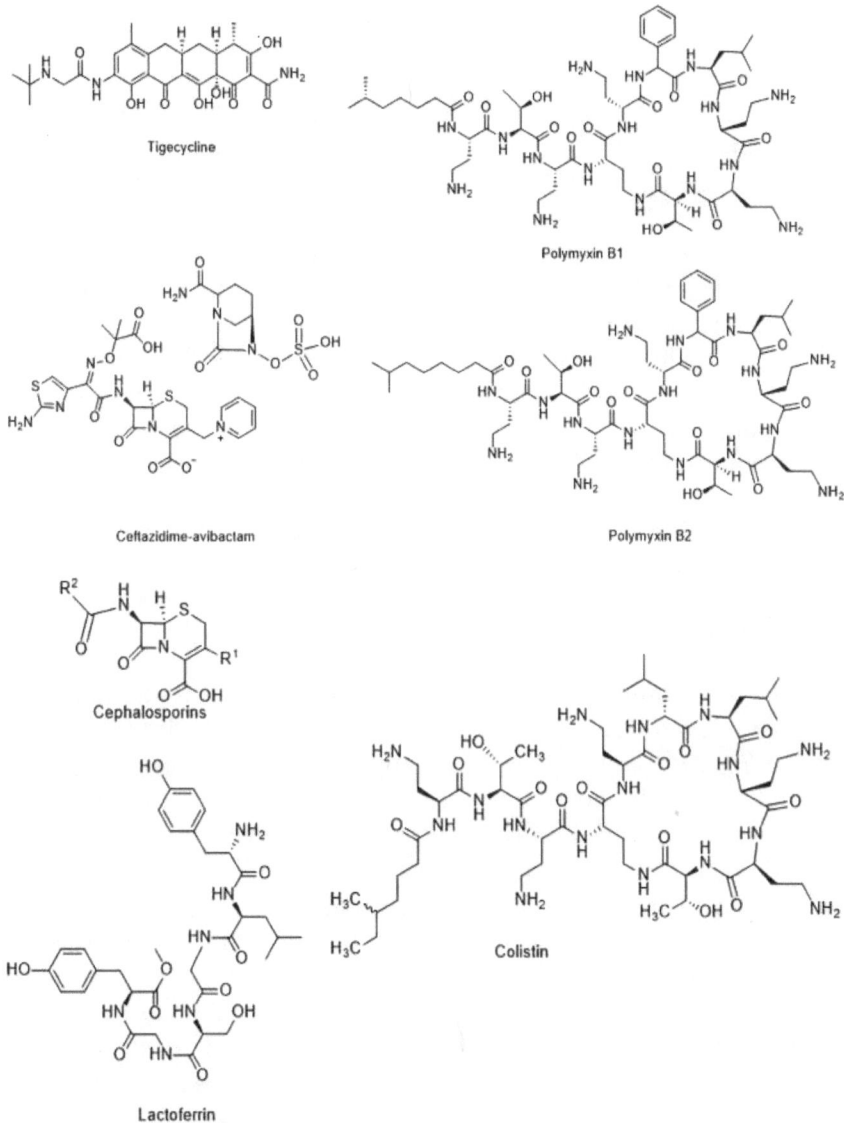

Fig. (13). Chemical structures of Tigecycline, Polymyxin B, Ceftazidime-Avbactam, Cephalosporins, Colistin, and Lactoferrin.

Tigecycline: This antibiotic of the tetracycline family is administered intravenously to treat serious infections caused by Gram-negative bacteria, including *P. aeruginosa*.

Ceftazidime-avibactam. It is an intravenously administered combination of antibiotics, including a β-lactamase inhibitor and a third-generation cephalosporin. It is effective against many antibiotic-resistant strains of *P. aeruginosa*.

Polymyxin B: This antibiotic treats infections caused by antibiotic-resistant *P. aeruginosa* strains. Like colistin, it can be toxic to the kidneys. Various treatments can be used to combat antibiotic-resistant *P. aeruginosa* infections besides antibiotics. These include phage therapy, immunotherapy, and antimicrobial peptides. Antimicrobial peptides (AMPs) are small molecules that occur naturally in many organisms, including plants, animals, and bacteria, and are used to defend against microbial pathogens. AMPs are effective against a wide range of microorganisms, including *Pseudomonas aeruginosa*. AMPs can still be effective against bacteria, even after antibiotics have failed. They target the bacterial membrane, making it harder for bacteria to develop resistance. LL-37, colistin, and lactoferrin are AMPs effective against *P. aeruginosa* [88].

Colistin is a polymyxin-type antibiotic often used as a last resort for treating infections caused by antibiotic-resistant bacteria, including *P. aeruginosa*. However, it is essential to note that colistin can be toxic to the kidneys and should be used with caution.

Lactoferrin is an antimicrobial peptide in various bodily fluids, including breast milk, tears, and saliva. Lactoferrin has been found to have antimicrobial activity against *Pseudomonas aeruginosa*, both by sequestering iron and by directly disrupting the bacterial membrane. Also, it has been shown to inhibit the growth of *P. aeruginosa,* biofilm formation, and enhance the activity of other antimicrobial agents, such as antibiotics and other AMPs. Lactoferrin fights against microorganisms and has properties that can help reduce the severity of infections caused by *P. aeruginosa* by regulating the immune system and reducing inflammation. However, lactoferrin may not be effective against all types of drug-resistant bacteria. Therefore, it is essential to use other strategies, such as improved infection control and responsible use of antibiotics, to combat the rise of drug-resistant bacteria. While lactoferrin shows promise as a potential therapy for treating infections caused by *Pseudomonas aeruginosa*, more research is necessary to determine its safety and effectiveness in clinical settings [88].

COMBINATION ANTIMICROBIAL AS THERAPY AGAINST RESISTANCE-DRUG BACTERIA

Research has conclusively proven that antibiotics targeting multiple molecular targets or complex macromolecular structures are significantly more effective than those targeting a single protein. This practical activity is because the interdependent subunits comprising enzymatic complex structures are not easily affected by a single but several genetic mutations. Synthetic sulfonamides, the first antibiotics discovered in the 1930s, have become ineffective because of the resistance caused by targeting a single metabolic enzyme, the dihydropteroate synthase. Resistance occurs through point mutations or lateral gene transfer. However, β-lactams, aminoglycosides, and tetracyclines, discovered later, were found to be more resistant to genetic mutations in their molecular targets [89].

Developing synthetic compounds that can effectively target multiple factors has proven to be a significant challenge. This has resulted in a poor track record in discovering and developing antibiotics. As a result, the focus has shifted towards natural sources for new antibiotics, potentially leading to improved outcomes. However, identifying potential compounds through traditional growth inhibition screens can be highly challenging, as there is a risk of discovering known antibiotic scaffolds. Innovative approaches, such as post-assay dereplication using resistance profiles or mass-spectrometry-based metabolomic fingerprinting, have shown promise but require significant resources. In light of these challenges, the use of combination therapy, which involves the use of distinct compounds, has emerged as a viable alternative [90].

Antibiotic combinations were initially introduced in the 1950s and 1960s for economic reasons. Today, detailed mechanical, clinical, and epidemiological data strongly support using combinations as the first-line treatment option. However, it is crucial to note that combinations of fixed-dose antibacterial drugs are rare in drug forms, except β-lactam-β-lactamase inhibitors. Despite this, antibiotic combinations in clinical settings are standard practice. Given these circumstances, it is imperative to reconsider the use of monotherapy as the standard and explore the potential of combination therapies. This shift in focus will enable us to target multiple points and reduce the likelihood of spontaneous resistance [89].

Extensively drug-resistant (XDR) and pan-resistant strains of bacteria are impervious to single-antibiotic treatment. Some experts advocate for the use of antibiotic combinations to enhance treatment efficacy and thwart the emergence of further resistance. However, there has been a continuing dispute among experts over whether this method can effectively forestall the development of resistance in typical bacterial pathogens like *Mycobacterium tuberculosis* (TB).

Due to *Mycobacterium tuberculosis* having shown antibiotic resistance, the worldwide prevalence of tuberculosis illness is increasing. The antibiotic resistance of *M. tuberculosis* includes the emergence of monoresistant, multidrug-resistant (MDR), extensively drug-resistant (XDR), and now drug-resistant (TDR) forms of the disease. *Mycobacterium tuberculosis*, commonly known as TB, and its drug-resistant strains continue to exist in some parts of the world due to three primary factors. Firstly, the local healthcare agencies in these regions failed to take immediate and effective measures to tackle the issue. Secondly, physicians did not receive additional training or experience to treat patients in ways that prevent resistance from developing. Finally, a lack of public awareness left many individuals unaware of the disease's severity and ease of transmission.

Thioridazine (TDZ), which is a derivative of chlorpromazine, has been proven to be an effective treatment for tuberculosis when combined with XDR-TB-resistant antibiotics. It enhances the ability of non-killing macrophages to destroy intracellular *M. tuberculosis*, which prevents the bacteria from expelling antibiotics before they reach their targets. It also reduces the activity of efflux pumps that contribute to *M. tuberculosis'* multidrug-resistant properties. TDZ has also effectively killed dormant *M. tuberculosis*, making it a promising candidate for developing an antitubercular drug to cure latent TB infections. Although TDZ may cause rare side effects, the doses used for treating XDR-TB patients are relatively low (200 mg/day) compared to those used for treating chronic psychosis (1000 mg/day). TDZ can be a safe alternative for treating MDR/XDR-TB and TDR-TB infections. Cardiac monitoring and precautions should be taken before and during short-term therapy (a few months) compared to the long-term treatment of psychosis. TDZ is considered a "salvage drug" and can improve the quality of life for XDR-TB patients. Based on current information, TDZ seems to be the best option for treating XDR-TB and potentially TDR-TB patients [91].

Data suggests that combination regimens can reduce the emergence of resistance *in vivo*, particularly for Gram-negative bacterial infections. However, in systematic reviews for *Acinetobacter*, combinatorial therapy seems ineffective in preventing the emergence of resistance. So, based on the heterogeneity of the data, there are no general conclusions about therapeutic results in this case. Furthermore, a recent case from Greece of XDR *A. baumannii* infections also found no clinical advantage of colistin-based combination regimens compared to colistin alone.

Nevertheless, a study thoroughly examined the effectiveness of combining colistin and rifampin in killing *A. baumannii* and preventing the emergence of colistin resistance in MDR *A. baumannii* [92].

Researchers conducted a study to observe the effects of three different medication combinations on two strains of *A. baumannii*. The experiment was conducted over 72 hours, during which the bacteria were exposed to varying levels of colistin and rifampin. The response of the bacteria was measured through regular bacterial counts. One strain was susceptible to colistin, while the other was resistant. Population analysis profiles were used to assess the emergence of colistin resistance. The results showed that medication combinations led to significantly greater killing of both isolates at low inoculum. Combinations containing 2 and 5 mg/L colistin also increased killing at high inoculum. Additionally, combinations were found to be additive or synergistic at various time intervals with all colistin concentrations against both isolates. The combination of colistin and rifampin successfully prevented the emergence of colistin-resistant subpopulations in the colistin-susceptible isolate, regardless of inoculum size. These findings are essential in improving the effectiveness of colistin-rifampin combinations against both colistin-susceptible and colistin-resistant MDR *A. baumannii* [92].

Selection pressure has led to carbapenemases in certain bacteria, which break down antibiotics like imipenem, meropenem, ertapenem, and doripenem. To address this problem, a combination of a β-lactamase inhibitor and carbapenem-type antibiotics is commonly used. Effective carbapenemase inhibitory medications, such as Avibactam, Vaborbactam, and Relebactam, can combat resistant bacteria (Fig. **14**).

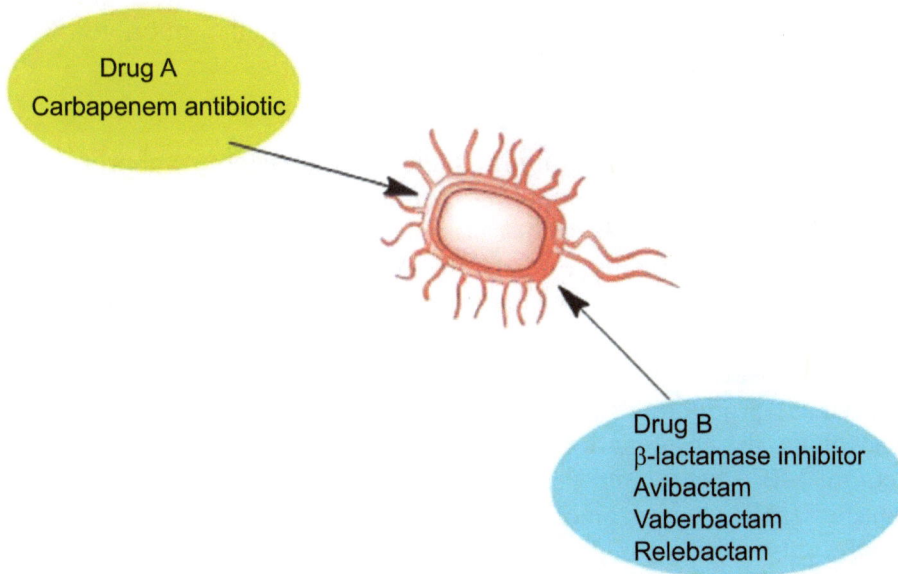

Drug A
Carbapenem antibiotic

Drug B
β-lactamase inhibitor
Avibactam
Vaberbactam
Relebactam

Fig. (14). One of the best combinations to combat drug-resistant bacteria.

The incidence of infections caused by Metallo β-Lactamases (MβL)-producing *Enterobacterales* and *Pseudomonas* is rising globally. These infections have alarmingly high mortality rates of over 30%. However, there is currently no universally accepted standard treatment for such diseases. It's worth noting that Aztreonam, an older β-lactam antibiotic, continues to be effective as MβLs cannot hydrolyze it.

It was observed that many strains of bacteria that produce MβL also produce enzymes that can break down aztreonam, a β-lactam antibiotic. Therefore, it is recommended to use avibactam, a potent β-lactamase inhibitor, in combination with aztreonam. A systematic review of 35 *in vitro* and 18 *in vivo* studies using aztreonam + avibactam for infections caused by MβL-producing Gram-negative bacteria has been conducted. The *in vitro* data showed that the combination of aztreonam and avibactam had high antimicrobial activity against 80% of MβL-producing *Enterobacterales*, 85% of *Stenotrophomonas*, and 6% of MβL-producing *Pseudomonas*, with a MIC of ≤ 4 mg/L [93].

Combination Ceftazidime-avibactam with Aztreonam

In a study of 94 patients, 83% were diagnosed with bloodstream infections. Fortunately, after treatment with aztreonam (ATM) + ceftazidime/avibactam (ATM/AVI), 80% of these patients could clinically recover within 30 days. Unfortunately, 19% did not survive. While new antimicrobials are available, aztreonam + avibactam (AVI) is a promising treatment option for Metallo--lactamases (MBL)-producing bacteria, particularly *Enterobacterales,* although it may be less effective against *Pseudomonas* [94].

The use of β-lactamase inhibitors such as avibactam, vaborbactam, and relebactam has dramatically improved patient outcomes in the United States. These inhibitors effectively block the activity of *Klebsiella pneumoniae* carbapenemases (KPCs), thus restoring the effectiveness of their partner β-lactam (BL) against KPC-producing pathogens. Ceftazidime, meropenem, and imipenem-cilastatin are the respective partner β-lactams. Metallo-β-lactamases (MBLs) are considerably more abundant than KPC in several provinces among carbapenemase-producing *Enterobacteriaceae* (CPE) worldwide. However, in some hospitals, the combination of ceftazidime-avibactam used to combat KPCs has led to MBL becoming the most common carbapenemase. Metallo-bet--lactamases (MBLs) use zinc ions to break down beta-lactam antibiotics. Lactamase inhibitors cannot stop them (BLIs are commercially available). Treating infections caused by MBL-producing E*nterobacteriaceae* is a significant and ongoing medical challenge.

Aztreonam, the only monobactam currently used in medicine, is not broken down by MBLs due to weak enzyme binding. However, most MBL-producing strains also produce extended-spectrum β-lactamase (ESBL), AmpC beta-lactamase, or other beta-lactamases that break down aztreonam. Recent research has explored the use of beta-lactamase inhibitors (BL/BLIs) in combination with aztreonam to address this issue. With this approach, a BL/BLI like ceftazidime-avibactam can stop ESBL and AmpC beta-lactamases, while aztreonam can avoid MBL-mediated breakdown and kill bacteria. This combination has shown high effectiveness against NDM-producing *Enterobacteriaceae* and *Stenotrophomonas maltophilia* (Fig. **15**).

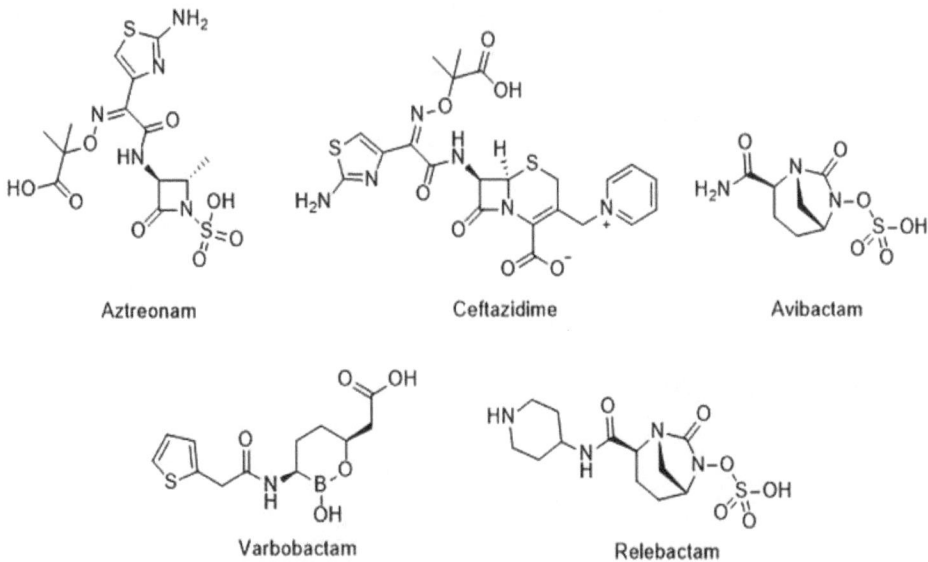

Aztreonam   Ceftazidime   Avibactam

Varbobactam   Relebactam

Fig. (15). Chemical structures of the β-lactam antibiotics Aztreonam Ceftazidime and the β-lactamases inhibitors Avibactam, Vaborbactam, and Relebactam.

Aztreonam and ceftazidime-avibactam are increasingly used together in clinical settings due to their effectiveness in laboratory tests and the lack of alternative treatments. In Spain, ten patients (including five with bacteremia) contaminated with *K. pneumoniae* assembling three β-lactamases, NDM-1, OXA-48, and CTX-M-15, were treated with this combination, and six of the ten patients were cured within 30 days.

A neutropenic child was reported to have bacteremia due to carbapenem-resistant *Enterobacter hormaechei* but was successfully treated with combination therapy using aztreonam and ceftazidime-avibactam. The strain resisted ceftazidime-avibactam and meropenem-vaborbactam combinations but was identified as having blaKPC and blaNDM, later confirmed as KPC-4 and NDM-1. *In vitro* testing confirmed synergy between the two agents, and steady-state therapeutic drug monitoring was performed, maintaining free drug concentrations above 2 μg/mL for aztreonam and ceftazidime and above 2.5 μg/mL for avibactam for about 50% of the dosing intermission. The patient acquired a 2-week therapy system that resulted in a clinical and microbiologic cure.

However, several significant problems need to be addressed regarding this issue. Relying solely on new BL/BLI combinations to treat CPE mediated by KPC enzymes may not always be practical. It is crucial to have knowledge of the local epidemiology, conduct vulnerability testing, and perform prompt diagnostic examinations to select the appropriate carbapenemase classes. A variety of molecular and immune-chromatographic tests have become available [95]. Additionally, avibactam has been found to protect aztreonam from hydrolysis caused by non-MBL β-lactamases, such as ESBL and AmpC. Therefore, it is logical to administer ceftazidime-avibactam initially, followed by aztreonam. Nevertheless, with the brief half-life of avibactam and the conditions for prolonged intake in the mix with ceftazidime (≥ 2 hours), apprehension has been raised that the available avibactam engagements may not be satisfactory. It is also essential to determine the right BLI concentration in the combination.

Despite the valuable insights obtained from the study regarding the efficacy of the potential dosing regimen for the combination, it is crucial to emphasize that the pharmacokinetic (PK) characteristics of β-lactamase inhibitors (BLIs) have not been adequately characterized in patient populations at the highest risk for suboptimal exposures. These include critically ill patients receiving renal replacement therapy. It is imperative to note that BL/BLI agents, including ceftazidime-avibactam, are solely available in fixed-dose combinations, severely limiting the scope for dosage adjustments in case of modified medicine elimination or raised lowest inhibitory concentrations vs mark pathogens. Consequently, it is crucial to ensure that exposures to both BLs and BLIs are consistent and adequate to achieve the expected Therefore, it is essential to ensure that the levels of both β-lactam antibiotics (BLs) and beta-lactamase inhibitors (BLIs) are consistent and adequate to achieve the intended effects in patients with various pathogens. It is necessary to conduct additional research to examine the pharmacokinetic/pharmacodynamic (PK/PD) exposures of patients receiving aztreonam and ceftazidime-avibactam, even if initial treatments have been successful. Immediate attention and in-depth studies are required for prolonged or continuous infusions of one or both agents. On the other hand, individualized BLI treatments can be tailored to each patient and pathogen without the risk of unnecessary BL dispensing. *In vitro*, the mix of available BLs or BLIs with aztreonam has been studied. *In vitro*, the assortment of aztreonam with other commercially obtainable BLs or BLIs has been completed. Still, dynamic prototype systems or patient studies must be pursued for more insights. While well-controlled, randomized studies in these areas would be beneficial, they could be impractical. As the use of aztreonam combination therapies increases, it may be more practical to create a registry of patient cases that includes detailed pharmacokinetics assessments. This case outlines a valuable treatment approach for infections caused by Gram-negative bacteria that produce metallo--lactamases, which are currently difficult to treat. The combination of ceftazidime-avibactam and aztreonam demonstrates clinical utility. It could delay the surge until agents with potent *in vitro* activity against MBLs become available to patients [94].

The prevalence of *Klebsiella pneumoniae* carbapenemase (KPC) versions, specifically KPC-2 and KPC-3, is a global concern. These versions resist most β-lactams, including carbapenems, but can be treated with new β-lactam/--lactamase inhibitors like ceftazidime-avibactam. However, it is alarming that numerous KPC variants resistant to ceftazidime-avibactam have been discovered in isolates obtained from clinical samples or experimental studies after its introduction. These variants have various mutations in specific areas, including point mutations, insertions, and deletions. It is crucial to understand the impact of these mutations on treatment and monitor the development of KPC variants and their resistance to ceftazidime-avibactam over time and space. Therefore, several studies comprehensively examine the mutational landscape of KPC β-lactamase and its resistance to ceftazidime-avibactam using a multidisciplinary approach, including epidemiology, microbiology, enzymology, and thermodynamics.

Research has shown that resistance is linked to three specific areas where insertions and deletions occur more than in other β-lactamases. However, becoming resistant to ceftazidime-avibactam can result in a weaker resistance to other β-lactams and decreased enzyme stability. Despite this, the natural stability of KPC may explain why it is prone to developing mutations that provide resistance to ceftazidime-avibactam, often through insertions and deletions within the enzyme [95].

Sulbactam-durlobactam

Acinetobacter baumannii has been identified as a critical priority for new treatment options. Currently, 33 accepted genospecies of *Acinetobacter* have been defined by DNA-DNA hybridization. Among those related to human disease are *Acinetobacter calcoaceticus, A. baumannii*, A. haemolyticus, A. junii, A. lwoffii, A. johnsonii, and *A. ursingii*. The species *A. baumannii*-calcoaceticus complex (ABC) has been associated with severe infections. This complex presented four highly similar species that cannot be differentiated by phenotypic tests: *A. baumannii*, *A. pittii*, *A. nosocomialis* (most frequently isolated in hospital-acquired infections), and *A. calcoaceticus*. The severe infections associated with ABC include hospital-acquired bacterial pneumonia/ventilator-associated bacterial pneumonia (HABP/VABP), complicated urinary tract infections (cUTIs), bloodstream infections, and wound infections. Approximately two-thirds of ABC infections are caused by multidrug-resistant (MDR) isolates. Severe infections caused by MDR ABC isolates have been reported to have high morbidity rates, and even the mortality rate can reach up to 50% or more. So, there is an urgent need to identify new antimicrobial agents to treat these infections. Sulbactam (SUL), a class A β-lactamase inhibitor (BLI), exhibits antibacterial activity against ABC isolates. However, its use in monotherapy is limited by the possible increase in resistance.

On the other hand, durlobactam, also known as ETX2514, is a novel β-lactamase inhibitor structurally distinct from sulbactam. Durlobactam can inhibit a broader range of β-lactamase enzymes than sulbactam, including class A, class C, and some class D enzymes. Durlobactam also penetrates bacterial cells and targets β - lactamases inside the cell, which is impossible with many other β-lactamase inhibitors. Together, sulbactam and durlobactam form a potent combination that can overcome many mechanisms of resistance used by Gram-negative bacteria, demonstrating activity against many Gram-negative pathogens, including carbapenem-resistant *Enterobacteriaceae* (CRE) and *Acinetobacter baumannii*. Sulbactam-durlobactam combination has been clinically proven effective and well-tolerated for treating complicated urinary tract, intra-abdominal infections, and hospital-acquired pneumonia caused by multidrug-resistant Gram-negative bacteria.

The FDA has approved the sulbactam-durlobactam combination (Fig. **16**) to treat patients specifically with bloodstream infections, hospital-acquired, or ventilator-associated bacterial pneumonia caused by *A. baumannii*-calcoaceticus complex (ABC) [96, 97].

Combination of Colistin, Rifampicin

As mentioned above, carbapenem-resistant *Acinetobacter baumannii* (CRAB) is a type of bacteria causing increasing infections in hospitals worldwide. Due to the high number of infections caused by multidrug-resistant *Acinetobacter baumannii*, it is possible to treat this species using combinations of older antimicrobial drugs. A study was conducted to observe and evaluate the effectiveness of intravenous and aerosolized colistin when combined with rifampicin in treating critical patients with nosocomial infections caused by multi-resistant *A. baumannii*.

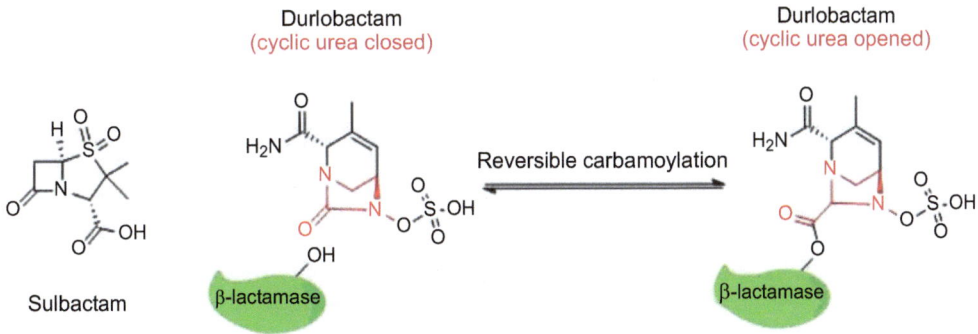

Fig. (16). Mechanism of action of Durlobactam.

In recent years, colistin (a mixture of cyclic polypeptides colistin 1 and 2) has resurfaced as a last alternative against pathogen infections such as *Pseudomonas aeruginosa, Acinetobacter baumannii* complex, and *Enterobacteria* producing carbapenemases. Previously, colistin fell into disuse due to its potential severe adverse effects, such as nephrotoxicity and neurotoxicity. It is an essential component of antibiotic treatments against highly resistant pathogens. The overuse of colistin can lead to resistance, which could be catastrophic. However, recent studies have found transferable plasmids between species containing the MCR-1 gene, which confers resistance to colistin. Rifampicin is a member of the class of rifamycins, which is a semisynthetic antibiotic derived from *Amycolatopsis rifamycinica*.

Clinical response to colistin + rifampicin was evaluated. The study involved 26 patients who were treated for different types of infections caused by *Acinetobacter baumannii*. Among them, 16 cases of nosocomial pneumonia were treated with aerosolized colistin along with intravenous rifampicin (10 mg/kg every 12 hours). In comparison, nine cases of bacteremia were treated with intravenous colistin (at a dosage of 2X106 IU three times a day) along with intravenous rifampicin (10 mg/kg every 12 hours). Three of these cases were associated with ventilator-associated pneumonia, and one case of nosocomial meningitis was treated using intrathecal colistin along with intravenous rifampicin. All the patients showed favorable clinical outcomes. However, three patients experienced moderate hepatic cytolysis, which is a known side effect of colistin combined with rifampicin. The need for a control group and the small number of patients limit this study. Therefore, the results should be interpreted with caution [98].

There is insufficient data on treating colistin-resistant *Acinetobacter baumannii* (CoRAB) infection, and nowadays, treating this condition represents a real challenge. However, some studies have shown effective combination therapies for CoRAB *in vitro* and *in vivo*. For example, a 100% synergistic effect of the colistin/rifampicin combination has been reported against nine clinical CoRAB isolates *in vitro*. However, a randomized clinical trial confirming the impact of this combination therapy has yet to be conducted [99].

No new emerging antibiotics have been found for colistin-resistant *Acinetobacter baumannii* (CoRAB). Many clinicians have tried various combinations of well-known antibiotics. Previous studies have shown that colistin/rifampicin combination therapy is an option for treating infection with CoRAB; however, these were *in vitro* or case studies. To date, there is no definite treatment for CoRAB. Nevertheless, colistin/rifampicin combination therapy may help achieve MR in some cases of pneumonia caused by CoRAB; however, its effectiveness in attaining CR appears doubtful. However, this regimen may help eradicate CoRAB, preventing the acquisition of further resistance and further spread of CoRAB. 'Partial synergy' of colistin and rifampicin can be a good prognostic factor for MR and CR in pneumonia caused by CoRAB [100].

Combination of Imipenem, Colistin, Sulbactam, and Tigecycline

A study has conclusively determined the effectiveness and interactions of imipenem, colistin, and tigecycline with older antibacterial agents against carbapenem-resistant *Acinetobacter baumannii*. Forty-three carbapenem-resistant *A. baumannii* isolates were collected from a hospital unit, and their imipenem, colistin, and tigecycline MICs were tested. An *in vitro* time-kill study was performed on eight randomly selected carbapenem-resistant isolates to evaluate the antibacterial activity of colistin, tigecycline, imipenem/sulbactam, and colistin/rifampicin combinations. The study found that colistin could kill the bacteria at 43 and 83 MIC concentrations, while tigecycline only showed the ability to inhibit bacterial growth at all concentrations. When combined, imipenem/sulbactam and colistin/rifampicin were most effective, demonstrating synergistic and bactericidal activity at a concentration of 13 MIC. The study also found that imipenem/sulbactam, colistin, and tigecycline were effective against carbapenem-resistant *A. baumannii* isolates. Although colistin is effective against carbapenem-resistant *A. baumannii*, for its specificity, the colistin/rifampicin combination is recommended [101].

Between 2006 and 2007, a study was conducted in Singapore to collect BA isolates from all public hospitals in the country. The MICs were determined using the recommendations of the Clinical and Laboratory Standards Institute (CLSI). A PCR-based method was used for genotyping all carbapenem-resistant *Acinetobacter baumannii* (CRAB) isolates, and their clonal relationship was established. Time-kill studies were conducted to test the effectiveness of polymyxin B, rifampicin, and tigecycline alone and in combination using clinically relevant unbound concentrations. A total of 31 CRAB isolates were identified, all of which were multidrug-resistant except for being susceptible to polymyxin B. Clonal typing revealed eight clonal groups, while 11 isolates showed clonal diversity. None of the drugs exhibited bactericidal activity at 24 hours when tested alone in the time-kill studies.

The analysis revealed that using polymyxin B alone may not be effective against polymyxin B-susceptible AB isolates. These results highlight that the *in vitro* synergy of antibiotic mixes in CRAB can vary depending on the strain. Also, these results could guide the choice of preventive therapy for CRAB infections [102].

Combination of Rifampicin/Imipenem

A pilot study was conducted from July 2000 to September 2001 with patients suffering from severe infections caused by carbapenem-resistant *A. baumannii*. Some patients were treated with a rifampicin/imipenem combination and monitored for changes in their cultures during and after treatment. The activity of rifampicin was also observed *in vitro*, and the genotype of the strains was determined using PFGE. Ten patients were selected, consisting of four patients diagnosed with ventilator-associated pneumonia and six patients suffering from other infections, including one catheter-related bacteremia and five surgical infections. Among these patients, three succumbed to their illnesses, and two were considered therapeutic failures. Still, due to the non-favorable pharmacokinetic profile of these antibiotics, attention needs to be paid to prevent resistance. The study suggests that rifampicin/imipenem may not be effective, and further studies are required to explore other rifampicin combinations [103].

Combinations of Antipseudomonal β-lactam (Piperacillin-tazobactam or Meropenem) with an Aminoglycoside (Gentamicin or Amikacin)

Research has shown that combining medications can be more effective than just one medication when treating *P. aeruginosa* infections. In a study involving 332 adult patients diagnosed with ventilator-associated pneumonia and found to have *P. aeruginosa* in their respiratory tract cultures, patients were randomly assigned to receive either combination therapy—an antipseudomonal β-lactam (piperacillin-tazobactam or meropenem) plus an aminoglycoside (gentamicin or amikacin)—or monotherapy with either an antipseudomonal β-lactam or an aminoglycoside.The study aimed to determine if patients experienced clinical cures by day 14, characterized by improving clinical signs and symptoms of pneumonia. The analysis found that combining therapies was more effective than just one therapy in achieving a clinical cure by day 14. The group using combination therapy had a clinical cure rate of 64.8%, while the monotherapy group had a rate of 47.9% (p=0.01). Additionally, the combination therapy group saw a higher rate of getting rid of the bacteria causing the illness and a lower mortality rate, although these differences were not statistically significant.

Therefore, treating *P. aeruginosa*-caused ventilator-associated pneumonia with a combination of an antipseudomonal β-lactam and an aminoglycoside can be more effective than using a single medication. However, this combination therapy can have harmful side effects, including kidney damage. It is essential to weigh the possible benefits and risks of combination therapy on a case-by-case basis. Factors such as the patient's characteristics, the severity of the infection, and the organism's antibiotic resistance patterns must be considered. Although combination therapy can generally be a practical approach to tackling *Pseudomonas aeruginosa* infections, the most effective strategy for treating antibiotic-resistant strains will depend on the strain's specific characteristics and the patient's circumstances [104]

Combination of Imipenem and Cilastatin/Relebactam

It has been documented that the imipenem/cilastatin/relebactam combination (Fig. 17), which is a combination of β-lactam and β-lactamase inhibitor, exhibits remarkable action versus carbapenem-resistant non-*Morganellaceae Enterobacterales* (CR-NME) and difficult-to-treat (DTR) *P. aeruginosa*. It is noteworthy that relebactam, although not an antimicrobial agent, effectively inhibits many Ambler class A and C enzymes.

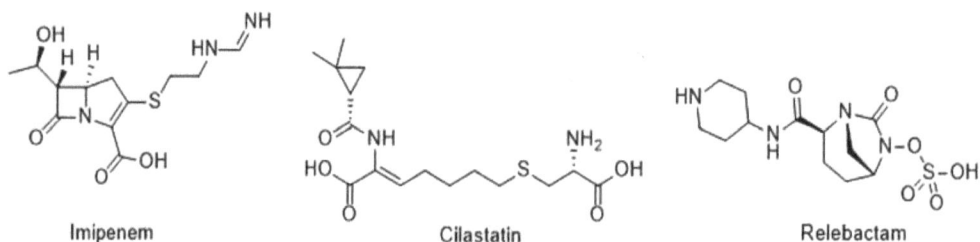

Fig. (17). Chemical structure of Imipenem, Cilastatin, and Relebactam.

This combination has been approved for the treatment of hospital-acquired bacterial pneumonia (HABP/VABP), ventilator-associated bacterial pneumonia (HABP/VABP), problematic urinary tract infections (cUTIs), and complex intra-abdominal infections (cIAIs) where restricted or no alternative therapies are known. The available data support its efficacy in treating HABP/VABP, cUTIs, and cIAIs caused by CR-NME and DTR *P. aeruginosa*.

The available data suggest that imipenem/relebactam (I/R) still has some effectiveness against *P. aeruginosa* isolates resistant to newer β-lactam antibiotics. Therefore, it is essential to perform susceptibility testing on all newer agents to identify the best treatment options for patients with highly resistant CR-NME and DTR *P. aeruginosa* infections. However, it is worth noting that further comparative studies are necessary to determine the most influential role of I/R, alone or in combination, for treating these infections. Until more data is available, I/R can still be considered a potential treatment for CR-NME and DTR *P. aeruginosa* infections, provided the benefits of its use outweigh the risks [105].

Ceftolozane and Tazobactam Combination

The ceftolozane/tazobactam (C/T) combination is used in the clinic to treat complicated urinary tract infections and intraabdominal infections in adults (Fig. **18**). Ceftolozane, an antibiotic of the β-lactam type, interferes with the construction of protective cell walls. Tazobactam is a drug that inhibits bacterial β-lactamases, enzymes that induce the breakdown of β-lactam antibiotics, such as ceftolozane, rendering the bacteria resistant to antibiotic activity. Therefore, tazobactam enables ceftolozane to combat bacteria otherwise impervious to its action.

Tazobactam   Ceftolozane

Fig. (18). Chemical structure of Tazobactam and Ceftolozane.

The mix of ceftolozane and tazobactam can be used as an initial treatment in characteristic clinical conditions even before an antibiotic-resistant *P. aeruginosa* isolate is noticed. These situations may include critically ill patients who are in the ICU setting who have failed therapy with a carbapenem and have a high probability of being infected with multidrug-resistant (MDR) *P. aeruginosa*. Also, patients with cystic fibrosis, those who have undergone lung transplants, and patients who have previously been infected or colonized with MDR *P. aeruginosa*. The ceftolozane/tazobactam combination was authorized for medical benefit in the USA in December 2014 and the EU in September 2015 [106].

Colistin/ciprofloxacin Combination

Persister cells are accountable for antibiotic therapy failure and the emergence of antibiotic resistance. A study investigated the lethal effects of antibiotic combinations on persister cells of *Pseudomonas aeruginosa* isolates. Colistin plus ciprofloxacin combination eradicated the *P. aeruginosa* persister cells. The study found that using both antibiotics simultaneously was more effective in eradicating bacteria than using them sequentially. Colistin persisters had the most significant reduction in intracellular ATP concentration. Moreover, the overexpression of the spoT gene was only found in colistin-persister cells; to the contrary, the relA gene was overexpressed in all persister cells, likened to untreated parent cells. According to the results of a Transmission Electron Microscopy (TEM) analysis, it is suggested that distinct mechanisms could generate persister cells depending on the antibiotic being used. Specifically, cell elongation and damage in the cell wall or membrane were observed in colistin persisters, while DNA condensation was identified in amikacin persisters. On the other hand, outer membrane vesicles were found in ciprofloxacin persisters [107].

Combinations of Colistin with other Drugs

Gentamicin alters combinations of colistin with fosfomycin or meropenem. Colistin has been studied alone or in various combinations in treating osteomyelitis caused by carbapenemase-producing *Klebsiella pneumoniae*. These infections are often caused by carbapenemase-producing *Enterobacteriaceae* (CPE) and are particularly challenging to treat due to biofilm and poor antibiotic diffusion in bone tissues. These infections are commonly seen in post-operative osteomyelitis and prosthetic joint infections. While ceftazidime/avibactam has been approved as a treatment option, other options are limited and may rely on older antibiotics with less efficacy.

In experiments conducted at 4% MIC, colistin alone showed rapid bactericidal activity for the first six hours, but regrowth occurred after nine hours of incubation. The MICs for the regrown isolates were like those of the initial strain.

Meropenem alone had slow bactericidal activity, decreasing the initial inoculum by only 2 log10 at nine hours. Fosfomycin and gentamicin exhibited earlier bactericidal action, but regrowth was observed. Gentamicin and colistin had rapid bactericidal activity, but regrowth occurred. Colistin with fosfomycin prevented regrowth. The most effective combination was colistin, meropenem, and gentamicin, with an immediate and complete bactericidal effect at three hours and no subsequent regrowth.

After testing, it was found that tigecycline alone did not completely kill the bacteria, and there was some regrowth after nine hours. When tigecycline was combined with colistin, the effectiveness of colistin alone was lost, and the combination worked against killing the bacteria. Additionally, when different concentrations of tigecycline were tested with a fixed concentration of colistin, it was observed that the combination's bactericidal activity decreased in a dose-dependent manner. However, the combination of colistin and tigecycline should be cautiously used as it may have an antagonistic effect.

It is not recommended to use colistin alone to treat osteomyelitis caused by KPC-producing *Enterobacteriaceae* due to its low effectiveness in the body and the rise of antibiotic-resistant strains. Combining colistin with meropenem, with or without gentamicin, could be helpful but may not entirely prevent the emergence of colistin-resistant strains. The combination of colistin and fosfomycin shows promise and requires further in vivo evaluation [108].

The combinations shown in Table **3** are just a tiny sample of the numerous reported. They were chosen to illustrate the different drugs that can be used. However, Table **3** shows that only three combinations belong to the β-lactam antibiotic/β-lactamase inhibitor type. Nevertheless, the table indicates that other combinations are as effective as the ideal one.

Table 3. The drug combinations against several infections are discussed in this review.

Combination	Drug A	Drug B	Infection
Ceftazidime-avibactam/Aztreonam	Carbapenemase inhibitor	β-lactam antibiotic	*Klebsiella pneumoniae* carbapenemases.
Sulbactam/durlobactam	β-lactamase inhibitor	β-lactamase inhibitor	*A. baumannii*-A. calcoaceticus complex
Colistin/rifampicin	Cyclic polypeptides	Macrocyclic antibiotic	Carbapenem-resistant *A. baumannii*
Imipenem-colistin/tigecycline	β-lactam antibiotic/cycle polypeptides	Tetracycline antibiotic	Carbapenem-resistant *A.baumannii*

(Table 3) cont.....

Combination	Drug A	Drug B	Infection
Rifamicin/imipenem	Macrocyclic antibiotic	β-lactam antibiotic	Carbapenem-resistan *A. baumannii*
Piperacillin-tazobactam or meropenem/gentamicin or amikacin	Antipseudomonal β-lactam antibiotic	Aminoglyco side	*P. aeruginosa* infections
Imipenem/cilastatin-relebactam	β-lactam antibiotic	Ambler class enzyme inhibitor	Carbapenem-resistant non-*Morganellaceae* Enterobacterales
Ceftolozane/tazobactam	β-lactam antibiotic	β-lactamases inhibitor	*P. aeruginosa* infections
Colistin/ciprofloxacin	Cyclic polypeptides	Fluoroquinolone	*P. aeruginosa* persister cell

HYBRIDS AGAINST DRUG-RESISTANT BACTERIA

Understanding how bacteria develop resistance through genetic mutations is paramount in combating drug-resistant bacteria. These mutations, which occur frequently due to their ability to adapt to aggressive environments, can cause significant changes in the bacteria's cellular structure, leading to resistance. Bacteria can also acquire new genes through transformation, transduction, and conjugation, allowing bacteria to share resistance genes [109].

With their various resistance mechanisms, bacteria pose a complex challenge in our quest to develop effective antibacterial treatments. These mechanisms include efflux systems and antibiotic-inactivating enzymes. They also reduce outer membrane permeability and form biofilms, a particularly intricate resistance mechanism. *Pseudomonas aeruginosa*, a dangerous pathogen, exhibits adaptive antibiotic resistance, further highlighting the issue's complexity [110].

In the comparison between prodrugs and hybrids, a prodrug is a molecule that is not pharmacologically active on its own but can be converted into active forms through enzymatic or chemical reactions in the body. The purpose of designing a prodrug is to modify the pharmacokinetic properties of the active drug, such as its bioavailability, absorption, and permeability, without altering its pharmacological activity.

Prodrugs fall into three categories. The first category is carrier-linked prodrugs, which involve attaching an active drug to a pro-moiety. The active drug is released *via* enzymatic or chemical reactions that remove the moiety. The second category is bio-precursor prodrugs, where the active drug is modified at the molecular level through oxidation or reduction reactions. The third category is double prodrugs, which link two biologically active drugs in a single molecule.

The two drugs are linked together and can be separated using various methods to release their components.

The chemical union between two or more drugs led to the formation of a new chemical compound, designed as hybrids that fall in the pro-drug classification. Hybrids, with their unique properties, hold immense potential in our fight against drug-resistant bacteria. They can be formed by directly joining molecules or using a molecular connector to bind active molecules through a covalent bond. This bond can either be cleavable or non-cleavable. A cleavable connector allows the hybrid to be enzymatically bio-transformed when it reaches the site of action, known as the hybrid prodrug approach. Meanwhile, a non-cleavable linker, known as the hybrid drug approach, remains intact for its time in the body. For example, the valine-citrulline linker in the DSTA4637S hybrid is cleavable, while the hybrid cefiderocol contains a non-cleavable linker.

Antibacterial medications targeting a single organism are the most commonly used. However, bacteria have shown the ability to resist these medications, rendering them ineffective [111]. To combat this, the practice of polypharmacology, or using medicines with multiple targets, has become more reliable in preventing or slowing resistance [112]. As a result, healthcare providers often prescribe multiple antibiotics simultaneously to prevent resistance, increase the range of effectiveness, and optimize the drug dosage. In combination therapy, an adjuvant that may be inactive is used with an antibiotic— a strategy called the antibiotic–adjuvant approach [113]. An adjuvant can facilitate antibiotics' entry into cells through various mechanisms, such as enhancing membrane permeability, inhibiting inactivation enzymes, or impeding active efflux of the antibiotic molecule [114].

The possibility of a post-antibiotic era is now a significant threat due to the fast spread of multidrug-resistant bacterial pathogens in hospital and community settings. These bacteria can resist various antibiotics through different mechanisms, such as gene mutations, enzymatic degradation, active efflux through porins, and other permeability barriers.

Various antibiotics have accelerated the transfer of genes that cause drug resistance in bacteria. This result is mainly due to the selective pressure exerted by different groups of antibiotics. Plasmids, integrons, and transposons facilitate the spread of these genes. Multidrug resistance is more common among Gram-negative bacteria, which makes the situation more critical than in the case of Gram-positive pathogens.

The Infectious Disease Society of America introduced the 10-20 program to address the dwindling supply of antibiotics. The objective was to create at least ten novel antimicrobial agents by 2020 that can combat bacterial infections. Currently, only six agents (ceftolozane-tazobactam, ceftaroline fosfate, ceftazidime-avibactam, meropenem-vaborbactam, delafloxacin, and secnidazole) have been developed and proven effective in treating drug-resistant Gram-negative bacterial infections.

Unfortunately, the limited supply of drugs makes it difficult to treat multidrug-resistant (MDR) Gram-negative bacterial infections, which continues to be a significant challenge. To overcome drug resistance in these organisms, it is crucial to expand our antimicrobial options or create new strategies for treatment and recovery [115].

Gram-negative bacteria's resistance to currently available antibiotics is mainly due to the overexpression of always-present efflux pumps and the protective outer membrane. Using a combination of two or more antibiotics could be a good approach. It would have more antimicrobial mechanisms and be less likely to cause resistance and mortality, leading to better clinical outcomes. However, *in vivo* studies did not demonstrate these advantages [115].

Researchers have developed antibiotic hybrids to tackle the growing problem of drug resistance and improve the efficacy of current antibiotics. These hybrids are synthesized by linking two molecules through a covalent bond to create a new, synthetic compound.

Combination therapies can be roughly classified by four principal modes of action by which compounds can enhance each other's activity.

   i. A second compound (an adjuvant) prevents the degradation or modification of the primary drug (an antibiotic).
  ii. A second compound (an adjuvant) allows the accumulation and retention of the primary drug (an antibiotic) by inhibiting the efflux pumps.
 iii. A second compound (an adjuvant) inhibits cells' intrinsic repair pathway or tolerance mechanism to the primary drug (an antibiotic).
  iv. A second compound is an antibiotic targeting a similar or different pathway inhibited by the first drug.

Implementing these strategies involves combining multiple antibiotics or pairing an antibiotic molecule with an adjuvant [116].

Cadazolid (Quinolone/Fluoroquinolone Hybrid)

Most studies on this topic have focused on fluoroquinolones and aminoglycoside molecules. These approaches have shown promising results in improving antibiotic access to the target and could lead to more effective treatments in the future [115].

Antibiotics of the synthetic type, oxazolidinones, effectively treat infections caused by Gram-positive bacteria. They bind to the ribosome's peptidyl transferase center (PTC), which is responsible for bacterial protein synthesis, and disrupt the peptidyl transferase reaction. Cadazolid (Fig. **19**) is the first antibiotic in the quinoxolidinone family that combines the pharmacophores of oxazolidinones and fluoroquinolones. It is being tested for its effectiveness in treating gastrointestinal infections caused by *Clostridium difficile*, which is common in hospitalized patients.

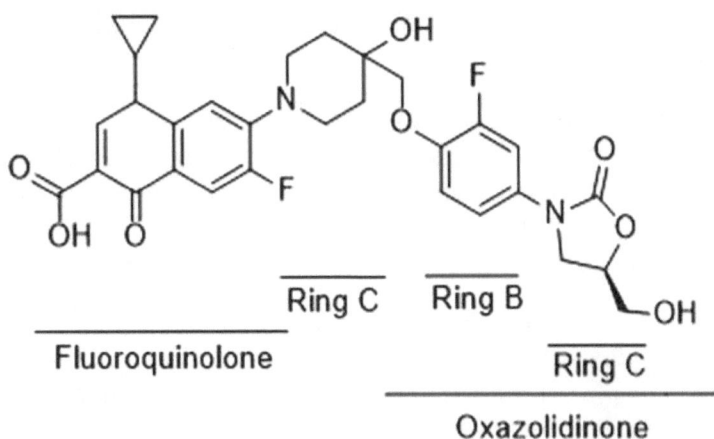

Fig. (19). Cadazolid.

This study utilized an isolate that exhibits resistance to the antibiotic linezolid, an oxazolidinone. The objective was to comprehensively understand cadazolid, another oxazolidinone antibiotic, and its mechanism of inhibiting bacterial growth. Cryo-electron microscopy was used to establish that cadazolid binds to *E. coli's* ribosome, disrupting tRNA binding to the A-site of the protein synthesis machinery. This interference ultimately restricts bacterial growth, making cadazolid a viable treatment option for linezolid-resistant strains [117].

Unlike traditional treatments, such as fidaxomicin, metronidazole, and vancomycin, Cadazolid boasts a success rate of up to 80%. This results because it targets the bacteria's protein synthesis, preventing them from producing toxins and spores. Even though Cadazolid underwent Phase III clinical trials, its development was discontinued after careful analysis of the data [118].

Tobramycin-based Hybrids As Adjuvants Potentiate Legacy Antibiotics

Tobramycin, an antibiotic, can boost the effectiveness of other antimicrobial agents against multidrug-resistant Gram-negative bacteria. This result means it can enhance the power of older antibiotics, giving them new life against resistant bacteria. Combining antibiotics in this way may offer a promising solution to the problem of drug resistance and help expand the range of available treatments. Developing antibiotics against *Escherichia coli*, *Acinetobacter baumannii*, *Klebsiella pneumoniae*, and *Pseudomonas aeruginosa* is challenging due to multiple resistance mechanisms. These include outer membrane impermeability, efflux pump overexpression, antibiotic-modifying enzymes, and gene and antibiotic target modification. One strategy to cope with these challenges is to use outer membrane permeabilizers that increase the intracellular concentration of antibiotics when combined.

One way to address antibiotic resistance is to rescue existing antibiotics or repurpose ones currently on the market. For example, researchers have found that combining two outer membrane-active components into a tobramycin-based hybrid antibiotic adjuvant can help antibiotics penetrate the outer membrane and accumulate more effectively within cells, making them more potent.

Researchers have expanded the concept of tobramycin-based hybrid antibiotic adjuvants by engineering up to three different antibiotic warheads, namely tobramycin, 1-(1-naphthyl methyl)-piperazine, ciprofloxacin, and cyclam, into a central 1,3,5-triazine scaffold. This was done to create tobramycin-based chimeras that are more effective against bacterial infections by targeting bacterial cell membranes.

This study found that Chimera 4 (TOB-TOB-CIP) (Fig. **20a**) effectively enhances the effectiveness of antibiotics against *P. aeruginosa*, including ciprofloxacin, levofloxacin, and moxifloxacin, even in cases of fluoroquinolone resistance. Moreover, the combination of a Chimera with ceftazidime/avibactam, aztreonam/avibactam, and imipenem/relebactam was able to reach the susceptibility breakpoints of ceftazidime, aztreonam, and imipenem, respectively, against β-lactamase-harboring *P. aeruginosa*. These findings highlight the potential of tobramycin-based chimeras as a new class of antibiotic potentiators that can effectively restore the activity of antibiotics against *P. aeruginosa*. The three-dimensional stereo structure of chimera 4 is illustrated in Fig. (**20b**) [119].

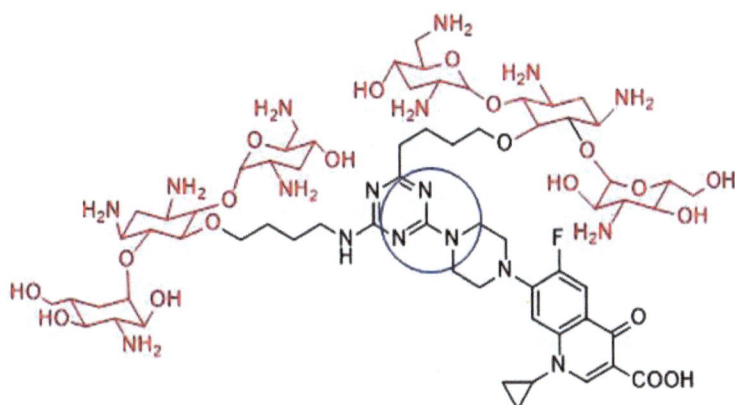

Fig. (20a). Chemical structure of Chimera 4 (TOB-TOB-CIP).

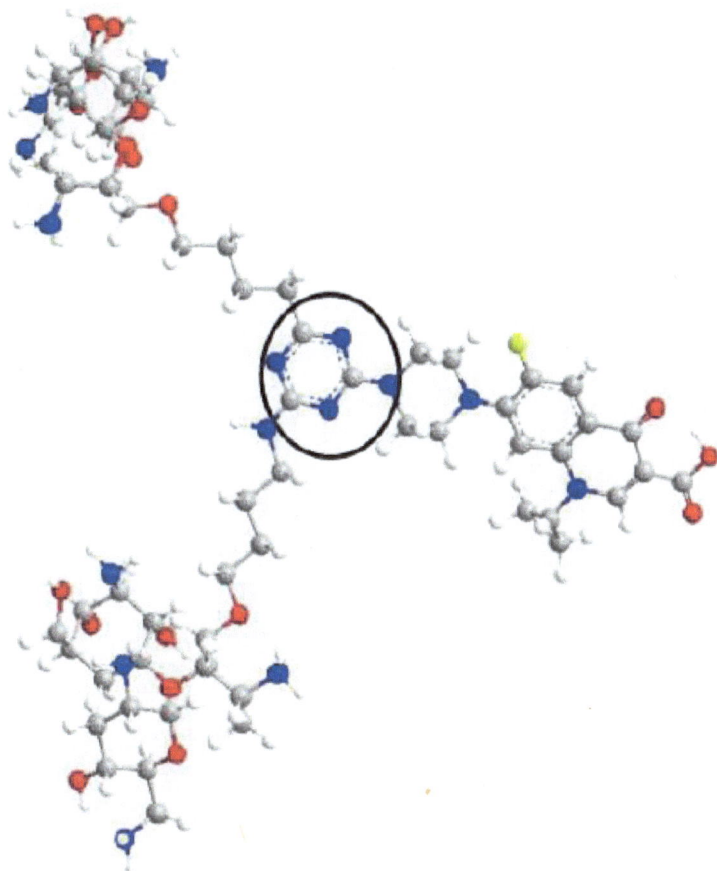

Fig. (20b). 3D structure of Chimera 4 (TOB-TOB-CIP) determined by Chem Draw 13.1 /Perkin Elmer) program.

Cephalosporin Hybridized with Vancomycin

Cefilavancin (TD-1792) (Fig. **21**) is a first-in-class vancomycin lipid II-cephalosporin heterodimer that exhibits bactericidal activity against Gram-positive pathogens. In a phase II trial, it was tested on patients with skin infections caused by Gram-positive bacteria. Patients between 18 and 65 received TD-1792 or vancomycin for 7-14 days. The trial was randomized and included an active control group. The results are shown in Table **4**.

Table 4. Bacteria activity of Cefilavancin (TD-1792) and Vancomycin.

-	TD-1792 Group	Vancomycin Group
Cure rates (n=170)	91.7%	90.7%
Microbiologically evaluable patients with MRSA at baseline (n = 75)	94.7%	91.9%
Microbiological eradication of Gram-positive pathogens (n = 126)	93.7%	92.1%

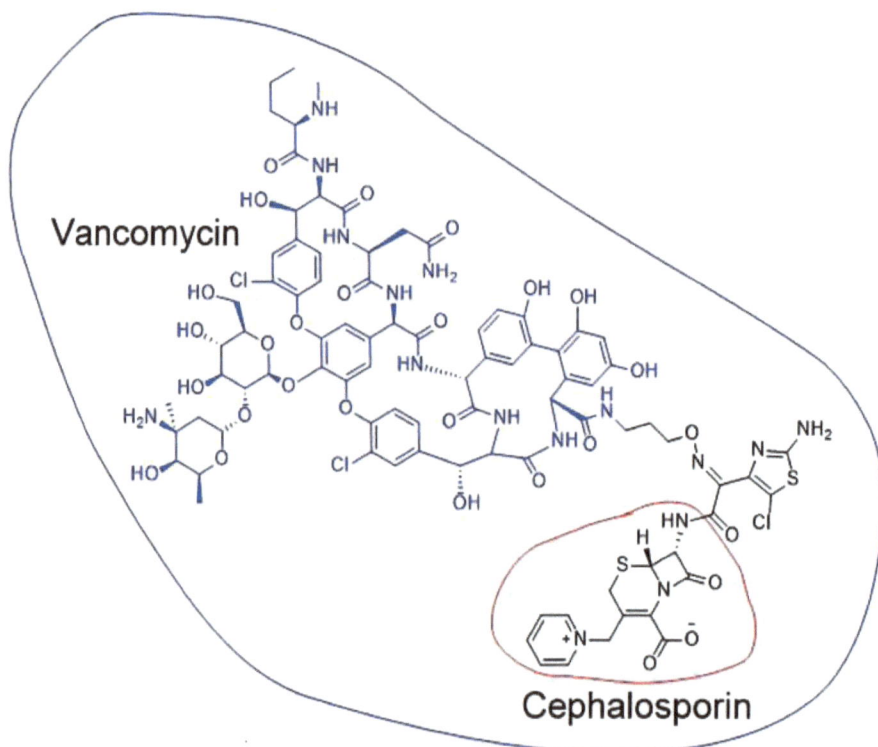

Fig. (21). Chemical structure of Cefilavancin (TD-1792) heterodimer.

During the study, seven patients had to stop taking the medication because of an adverse event. The types and severity of these events were similar in both groups, except for pruritus, which was more common in patients who took vancomycin. None of the patients who took TD-1792 had a severe adverse event. However, the findings show that cefilavancin and vancomycin have similar effects. Phase 3 studies are ongoing [120].

There is a significant medical need to treat prosthetic joint infections (PJIs) and other persistent bacterial infections. *Staphylococcus aureus* and coagulase-negative *staphylococci* (CoNS) are responsible for the majority of PJIs and are associated with higher treatment failure rates than observed with other causative pathogens. The treatment strategy for staphylococcal PJIs involves a prolonged rifampin-based combination therapy, including 2–6 weeks of intravenous antibiotic therapy in combination with oral rifampin, followed by rifampin plus a companion antibiotic for a total of 3–6 months. The formation of bacterial biofilms on the surface of the prosthesis appears to play an essential role in the pathogenesis of PJI. It is believed to be the underlying mechanism for persistence. It is well-known that pathogens living in bacterial biofilms can effectively evade host immune responses and tolerate antibiotic treatment.

Rifamycins are a powerful class of antibiotics that can effectively treat these infections. However, bacteria can become resistant to rifamycins quickly, so doctors typically only use them in combination with other antibiotics to minimize the risk of resistance developing.

A bifunctional molecule series based on rifamycin was designed, synthesized, and assessed to identify a dual-action drug. This drug would maintain the potent rifamycin activity against persistent pathogens while reducing antibiotic resistance and minimizing the development of rifamycin resistance. Scientists have been researching dual-acting antibacterial molecules that can target bacteria differently. One example is Ro 23-9424, combining a cephalosporin and a fluoroquinolone molecule.

Compound Ro 23-9424 has the characteristics of cephalosporin and does not function like fluoroquinolone. However, if it is chemically or metabolically broken down, the resulting fleroxacin component can act as an antibacterial agent in the same way as fluoroquinolone (Fig. **22**). Compound Ro 23-9424 is not a dual-acting antibacterial agent but a cephalosporin that works as a fluoroquinolone prodrug. Nevertheless, it is not more effective in preventing resistance development than either a cephalosporin or a fluoroquinolone.

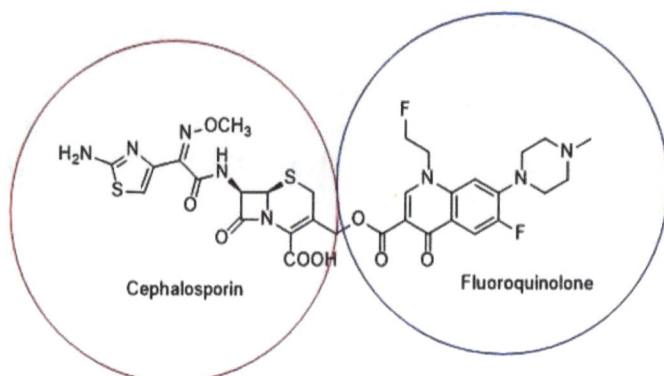

Fig. (22). Chemical structure of compound Ro 23-9424.

The development of rifamycin-based dual-acting antibacterial agents represents a promising strategy for treating persistent bacterial infections such as PJIs. The dual-action approach could address several significant issues associated with using rifamycin drug combinations and provide a better therapeutic solution in cases where the standard of care requires rifamycin combination therapy.

NP-2092 is an early-stage drug candidate for treating prosthetic joint infections. Hybrid 1(TNP-2092) is currently in clinical development and is the top molecule in the rifamycin-quinolone hybrid series. It is created by connecting a rifamycin and a quinolizidine core through the C-3 position on the rifamycin side and the C-8 position on the quinolizidine side. The linker group has a similar structure to the side chains found in rifampin and quinolizidine 4, which are joined together covalently (Fig. **23a**). The three-dimensional macrocyclic stereo structure of the rifampin component in hybrid 1(TNP-2092) is illustrated in Fig. (**23b**) [121].

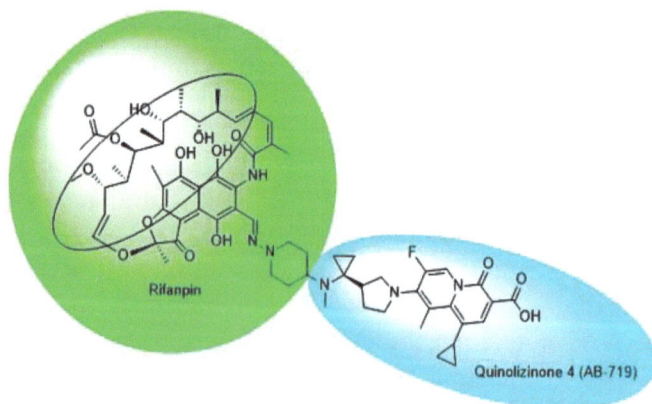

Fig. (23a). Chemical structure of the hybrid 1(TNP-2092).

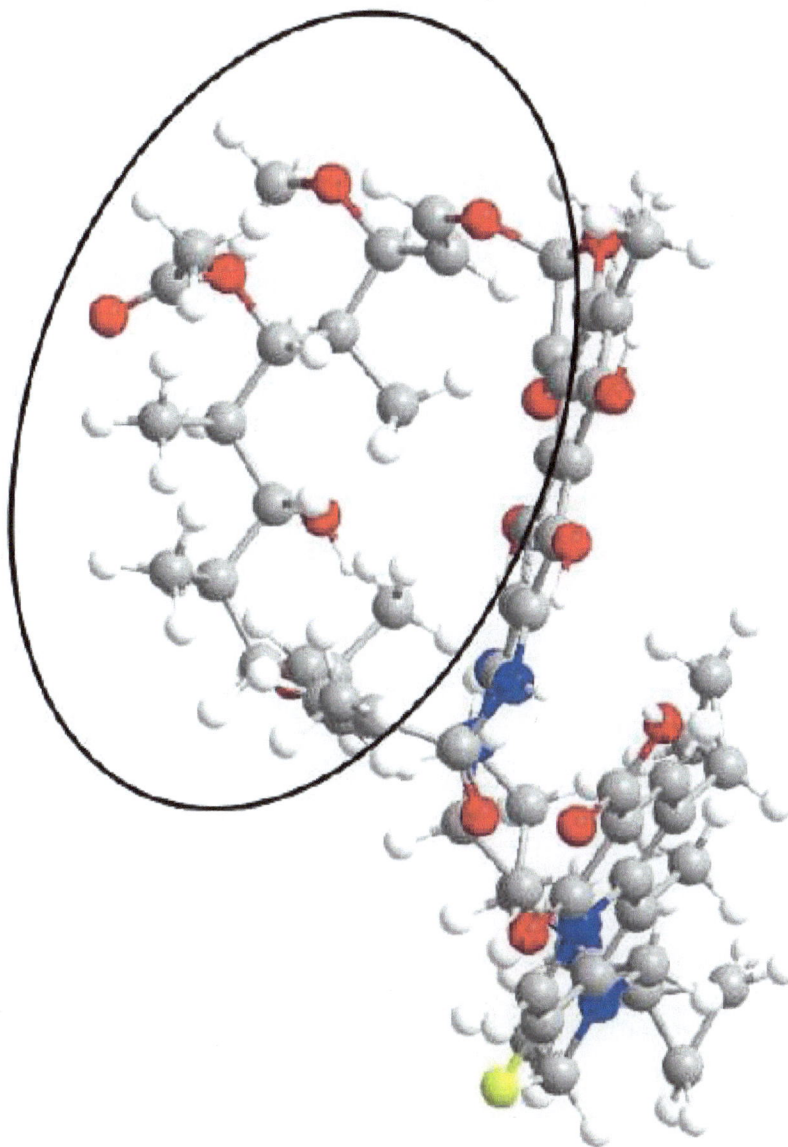

Fig. (23b). 3D structure of the hybrid 1(TNP-2092) determined by Chem Draw 13.1 /Perkin Elmer) program.

Antibacterial Flavonoid and Ciprofloxacin Hybrids

Twenty-one fluoroquinolone-flavonoid hybrids were synthesized using a well-planned pharmacophore model. These results helped develop a bacterial topoisomerase inhibitor that targets multiple areas.

Certain hybrid forms have demonstrated exceptional antibacterial properties against drug-resistant microorganisms. The combination of naringenin and ciprofloxacin has proven to be the most effective in treating infections caused by *Escherichia coli* ATCC 35218, *Bacillus subtilis* ATCC 6633, *Staphylococcus aureus* ATCC 25923, and *Candida albicans* ATCC 90873. Combining these two compounds has demonstrated 8-, 43-, 23-, and 88 times more significant activity than ciprofloxacin alone against the above bacterial and fungal strains.

This study has confirmed that two active analogs work in two ways by significantly impacting DNA gyrase and the efflux pump. Drug accumulation and DNA supercoiling assays were performed to demonstrate their efficacy. By introducing a flavonoid structure at the C-7 position of the fluoroquinolones, multiple beneficial interactions were augmented without compromising the binding mode of the fluoroquinolone moiety. These findings irrefutably imply the feasibility of developing antifungal drugs from fluoroquinolones with modifications at the C-7 position (Fig. **24**).

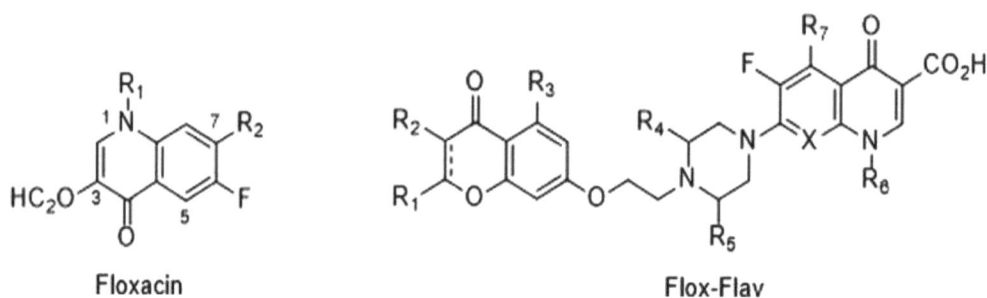

Fig. (24). General structures of a Fluoroquinolone and a Flox-Flav hybrid.

Through drug accumulation and DNA supercoiling assays, it was discovered that two active analogs possess a dual mode of action. This result can explain their unusual antibacterial activity against resistant strains and suggests that resistance development will be slow. Based on these findings, covalently binding an antibiotic with an efflux pump inhibitor can become a powerful weapon against resistant strains [122].

Through drug accumulation and DNA supercoiling assays, it was discovered that two active analogs possess a dual mode of action. This result can explain their unusual antibacterial activity against resistant strains and suggests that resistance development will be slow. Based on these findings, covalently binding an antibiotic with an efflux pump inhibitor can become a powerful weapon against resistant strains [122].

Antibiotics Hybridized with Siderophores

In 1987, a combination of siderophore and cephalosporin was synthesized (E-0702). This compound had antimicrobial properties against several Gram-negative bacteria, including *K. pneumonia, S. Typhimurium, S. marcescens, P. mirabilis*, and *P. aeruginosa.* Although resistant bacteria eventually emerged, this resistance was not caused by increased β-lactamase expression or reduced OmpF and OmpC porin proteins in the outer membrane. Instead, it was associated with a mutation in the tonB gene, which plays a vital role in siderophore-mediated iron uptake. The researchers also found that E-0702 was most effective against iron-starved bacteria but ineffective against iron-supplemented bacteria. It was hypothesized that E-0702 was incorporated into the bacteria through the tonB-dependent siderophore iron transport system.

There has been a growing interest in siderophore-antibiotic conjugates due to the rise of MDR Gram-negative bacteria. Researchers have recently developed a new compound, siderophore-β-lactam, effective against Gram-negative bacteria. It has been synthesized a set of monocarbam antibiotics conjugated with siderophore, and one of them, called MC-1, has shown promising results *in vitro* against MDR *P. aeruginosa*, ESBL-producing members of the Enterobacteriaceae and *Acinetobacter baumannii.*

More research is necessary to address this compound's hydrolytic instability. In the murine pulmonary infection model, the effectiveness of these compounds was limited due to their strong binding to plasma proteins.

Cefiderocol is a cephalosporin conjugate with structural features from ceftazidime and cefepime, which are 3rd and 4th-generation cephalosporins. Additionally, it has a catechol moiety at the C3 (Fig. **25**). This position allows cefiderocol to attach to ferric iron and enter cells through siderophore transporter proteins. Cefiderocol is effective against several types of Gram-negative bacteria, such as carbapenem-resistant Enterobacteriaceae, *P. aeruginosa*, and *A. baumannii*, both *in vitro* and *in vivo*. Its potency against these bacteria is due to the Trojan horse delivery strategy and its resistance to β-lactamase inactivation. Cefiderocol can combat OXA-58, NDM-, and IMP-producing *A. baumannii* isolates, *P. aeruginosa* isolates, and carbapenem-resistant Enterobacteriaceae, including KPC-, NDM-, IMP-, and VIM-producing isolates.

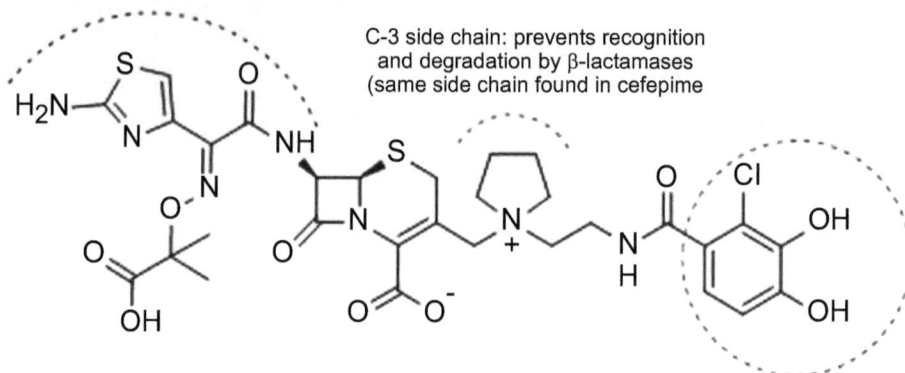

Fig. (25). Cefiderocol.

Clinical trials have already confirmed the ability of Cefiderocol to treat infections caused by common Gram-negative pathogens. In a prospective, multicenter, double-blind, non-inferiority trial, Cefiderocol 2000 mg was superior to imipenem-cilastatin 1000 mg three times daily for treating complicated urinary tract infections. The study enrolled 452 hospitalized adult patients, including immunocompromised patients.

Today, cefiderocol is available as Fetroja® (cefiderocol). This drug is indicated for patients 18 years of age or older for the treatment of complicated urinary tract infections (cUTIs), including pyelonephritis caused by the following susceptible Gram-negative microorganisms: *E. coli, K. pneumoniae, Proteus mirabilis, P. aeruginosa*, and *Enterobacter cloacae* complex [123, 124].

Antibiotics Hybridized with Novel Molecules

Researchers have discovered a new method for creating antibacterial agents. The technique involves combining indole through dimerization and carbazole hybridization. The process is accomplished through a simple one-pot reaction resulting in bisindole tetrahydrocarbazoles. These compounds are then oxidized to produce bisindole carbazoles with substitutions in the indole and carbazole scaffold (Fig. **26**).

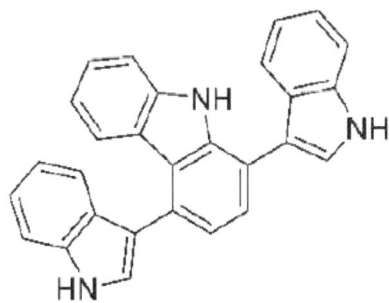

Bisindole carbazole   Bisindole tetrahydrocarbazole

Fig. (26). Chemical structure of Bisindole Tetrahydrocarbazoles and Bisindole Carbazole.

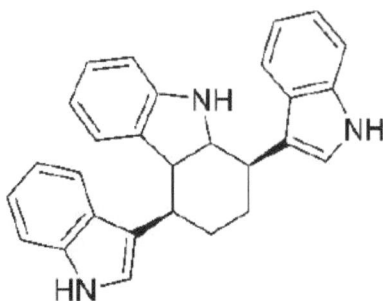

The effectiveness of different compounds was evaluated on several strains of *S. aureus*, including MRSA strains. Results showed that the most promising derivatives with 5-cyano substitution were based on MIC values. Surprisingly, tetrahydrocarbazoles were more effective than carbazole compounds and even more potent than standard antibiotics. These results indicate that these newly discovered compounds could be promising lead compounds for future studies [125].

There are countless ways to obtain hybrids, either with or without a linker. For instance, two antibiotics can combine, or an antibiotic can be paired with a new molecule, an enzyme inhibitor of bacteria, or a prodrug. The possibilities are endless. However, it is crucial that these hybrids are effective against drug-resistant bacteria and do not pose a risk to human health. Although some hybrids may successfully serve as antibiotic agents against drug-resistant bacteria, over time, bacteria may develop natural defenses against them due to selection pressure. The primary goal of hybrids is to provide a temporary solution while scientists search for better ways to combat these pathogens.

Resistance to the β-lactam antibiotics (penicillins, cephalosporins, and carbapenems) in pathogenic bacteria occurs most commonly through the production of β-lactamases that hydrolyze the β-lactam ring of these antibiotics, which is essential to their antimicrobial activity.

β-Lactamases use one of two distinct chemical mechanisms to achieve β-lactam ring opening (Fig. **27a**). From a Ser residue of the active site, a transient covalent bond is formed with the antibiotic, followed by hydrolysis of the ester associated with the enzyme, generating an inactive antibiotic. The other mechanism is by hydrolysis of β-lactam with a metal (Zn^2), which activates a water molecule, generating the hydrolytic species. Serine-β-lactamases, such as TEM, SHV, and

CTX-M have been the dominant enzymes in pathogens. However, metallo--lactamases (for example, NDM, VIM, and IMP) have been increasingly problematic in the clinic in the past few years.

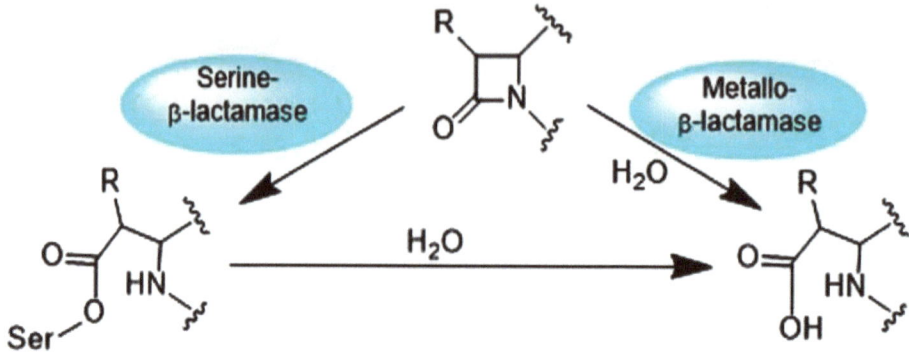

Fig. (27a). β-lactamases activity.

In 1976, scientists at Beecham Pharmaceuticals reported the synthesis of clavulanic acid, a new β-lactam-type compound with weak antibiotic activity but a potent inhibitor of serine-β-lactamases. Clavulanic acid was paired (Fig. **27b**) with amoxicillin in the first clinically syncretic β-lactam antibiotic-β-lactamase inhibitor combination, called Augmentin. Augmentin was a clinical and financial success, spurring the discovery of the penicillin sulfones serine-β-lactamase inhibitors tazobactam and sulbactam in the 1980s and more recent advances (see Fig. **27**, part b). In 2016, the FDA approved the first member of a new class of serine-β-lactamase inhibitors, avibactam (Fig. **27c**), a diazabicyclooctane with potent inhibition of many serine-β-lactamases that are poorly inhibited by existing drugs. A fixed-dose combination of avibactam with ceftazidime is available under the name Avycaz. Other diazabicyclooctanes are in various stages of development.

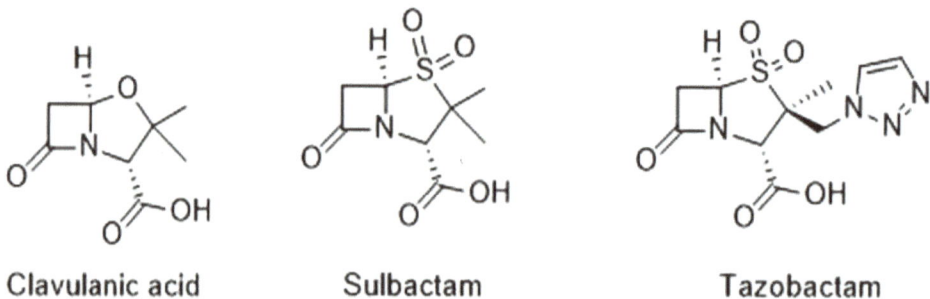

Clavulanic acid　　　　Sulbactam　　　　Tazobactam

Fig. (27b). Clavulanic acid, an inhibitor of serine-β-lactamases.

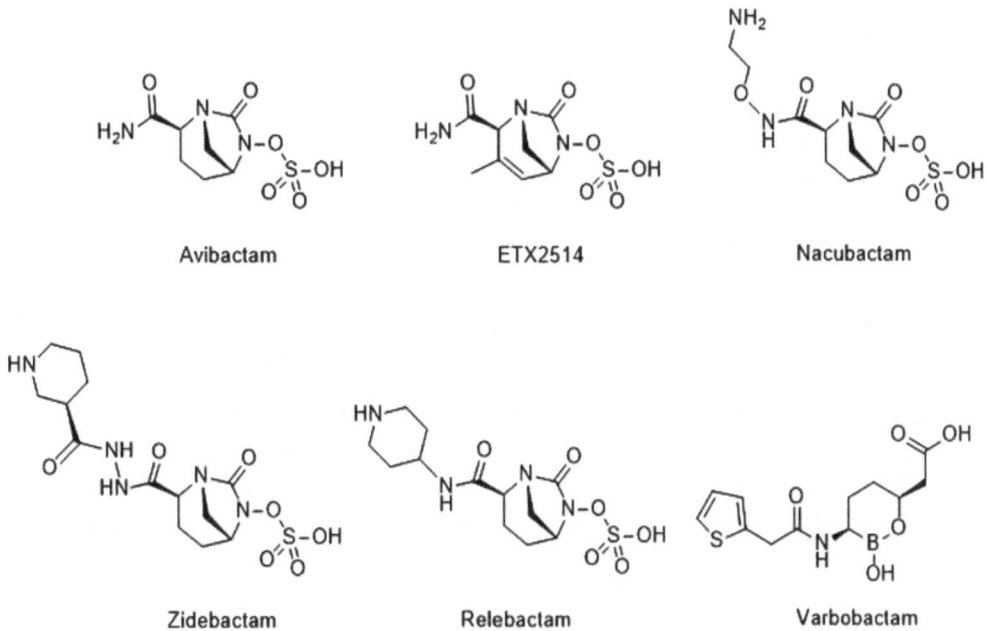

Avibactam   ETX2514   Nacubactam

Zidebactam   Relebactam   Varbobactam

Fig. (27c). Avibactam and another diazabicyclooctane derivative, a new class of serine-β-lactamase inhibitor.

In 2017, the FDA approved vaborbactam, a new cyclic boronate chemical scaffold with serine-β-lactamase inhibition. The combination of vaborbactam and meropenem is available under the trade name Vabomereh. Despite the growing importance of metallo-β-lactamases in the clinical failure of β-lactam therapy, no inhibitors of these enzymes are currently in late-stage clinical development.

QUORUM SENSING

The quorum-sensing system (QSS) regulates gene expression depending on the cell population density function. QSS bacteria produce and release chemical signal molecules called autoinducers (AI), whose concentration is a function of the cell density population. Detection of a minimum stimulating concentration of an AI alters gene expression. Therefore, QSS studies the communication between bacteria by producing and detecting AI. Several physiological actions are regulated by quorum quenching system (QQS) communication circuits in Gram-positive and Gram-negative bacteria. The various processes that microorganisms undergo include:

- Symbiosis (a close relationship with another organism for mutual benefit),
- Virulence (the ability to cause disease),

- Competence (the ability to take up DNA from the environment),
- Conjugation (the transfer of genetic material between two bacterial cells),
- Antibiotic production (the ability to produce compounds that kill other bacteria),
- Motility (the ability to move),
- Sporulation (the ability to form a dormant, resistant structure) and
- Biofilm formation (the ability to form a community of organisms that adhere to a surface).

In contrast to targeting vital processes of bacterial cells with antibiotics, 'anti-virulence' strategies aim to eliminate pathogenic features while maintaining cell viability, resulting in lower drug-induced selection pressure.

In *P. aeruginosa*, virulence factors are induced by AI produced by the quorum-sensing system (QSS). One type of autoinducer is an acyl-homoserine lactone-type (AHL), which many Gram-negative bacteria use. AHLs are synthesized by the bacteria and diffuse across the cell membrane. When the concentration of AHLs reaches a certain threshold, they go inside the cell, where they can bind to specific receptors in the cytoplasm. It triggers a response in the bacteria, such as the expression of genes involved in virulence or biofilm formation.

Another type of autoinducer is autoinducer-2 (AI-2), used by Gram-negative and Gram-positive bacteria. AI-2 is a furanosyl borate diester synthesized by the enzyme LuxS. Like AHLs, AI-2 can diffuse across the cell membrane and bind to receptors in the cytoplasm to trigger a response. In addition to AHLs and AI-2, other types of autoinducers are used by specific bacteria. For example, *Vibrio harveyi* uses the autoinducer-1 (AI-1), a fatty acid derivative.

In biofilm formation, bacteria can use quorum sensing to sense the presence of other bacteria in their environment and coordinate the expression of genes necessary for biofilm formation.

Understanding QSS and how it can be disrupted may lead to developing new strategies for controlling bacterial infections and improving the efficacy of antibiotics. Some of the most well-known examples of bacteria that use quorum sensing include *Vibrio fischeri, Pseudomonas aeruginosa*, and *Escherichia coli.*

Vibrio fischeri is a marine bacterium that produces light in response to quorum sensing, allowing it to communicate and coordinate its behavior with other bacteria.

Initially, N-(3-oxohexanoyl) homoserine lactone (3-O-HSL) is synthesized by LuxI and expelled from the cytoplasm. After that, a specific concentration is reached, and 3-O-HSL is introduced again in the cytoplasm and forms a transcription factor in combination with LuxR, activating the *Vibrio fischeri* luminescence genes (Fig. **28**) [126].

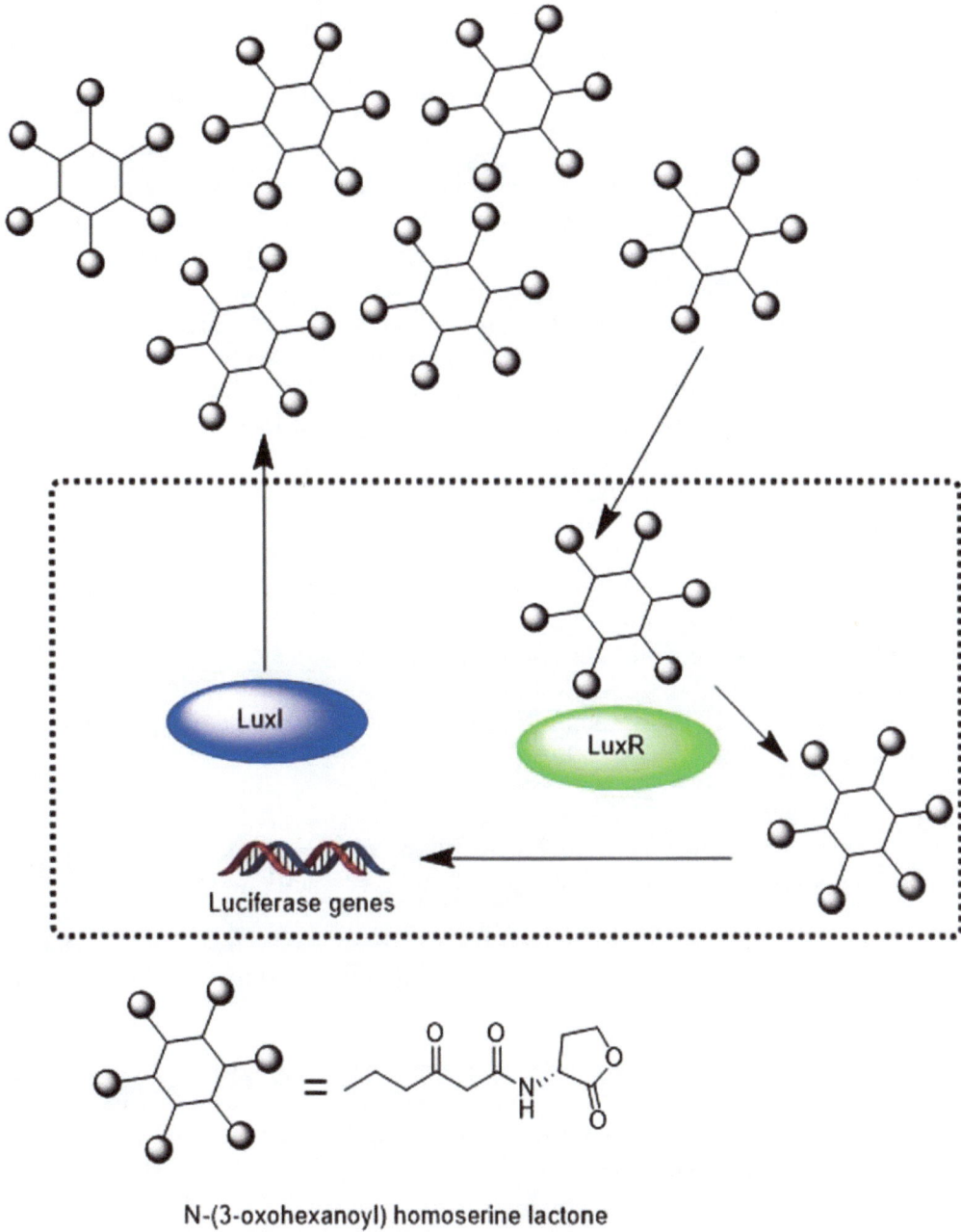

N-(3-oxohexanoyl) homoserine lactone

Fig. (28). *Vibrio fischeri* quorum sensing.

Pseudomonas aeruginosa can cause severe infections in humans, particularly those with weakened immune systems. Several Gram-negative bacteria control virulence factor production using LuxI/LuxR-type QSS circuits. For example, *P. aeruginosa* uses LasI/LasR and RhlI/RhlR circuits. Gram-negative bacteria, such as *Acinetobacter baumannii, Burkholderia cepacian*, and *Pseudomonas aeruginosa*, use N-acyl L-homoserine lactones (AHLs) as their primary QS autoinducer type.

Quorum sensing (QS) is a regulatory mechanism employed by *P. aeruginosa* to regulate hundreds of genes, among them many virulence factors, in response to population size. QS in *P. aeruginosa* involves four interconnected systems - *las*, *rhl*, *pqs*, and *iqs*. Initially, the *las* system was postulated as the most crucial since it was proposed that it controlled the other systems' functioning. In response to cell density, the LasR protein and its signaling molecule, 3-oxo-C12-HSL (odDHL), trigger the expression of *rhl*, *pqs*, and *iqs*. The N-butyryl-homoserine lactone (BHL)/RhlR complex, either alone or in coordination with other QS systems, activates the output of virulence elements, such as elastase B, rhamnolipid, hydrogen cyanide (HCN), pyocyanin, and the lectins LecA and LecB.

LasI and RhlI are enzymes that facilitate the synthesis of odDHL and BHL, respectively, using 3-oxo-acyl-ACP and crotonyl-ACP. After their synthesis, these AHLs are released into the environment and can be taken up by nearby bacteria through endocytosis (for odDHL) or passive diffusion (for BHL). Once inside, they bind to their corresponding cytoplasmic receptor proteins, LasR or RhlR.

The odDHL/LasR complex activates the expression of several genes, including *lasI*, *rhlI*, and *rhlR*. Similarly, the BHL/RhlR complex activates the expression of RhlI and other genes, some overlapping with the genes activated by LasR [127].

The *las* and *rhl* systems regulate the expression of approximately 300 genes in *P. aeruginosa,* which correspond to 4-12% of its genome and include the virulence genes.

The odDHL and BHL interact with a third system called the *Pseudomonas* quinolone signal quorum system (PQS). The communication between these systems is facilitated by 2-heptyl-3-hydroxy-4-quinolone (PQS) and its precursor, 2-heptyl-4-quinolone (HHQ). The PqsH monooxygenase converts HHQ to PQS, which is released extracellularly *via* exocytosis.

P. aeruginosa produces PQS, which is taken up by nearby bacteria through endocytosis. PQS then binds to the PqsR receptor, the Multiple Virulence-Factor

Regulator (MvfR). This binding activates the PqsR transcriptional regulator, increasing the PQS concentration in the bacteria's cytoplasm. The increased PQS concentration activates the expression of the pqsABCDE operon, leading to the production of enzymes PqsA, PqsB, PqsC, PqsD, and PqsE. These enzymes produce signaling molecules called 2-alkyl-4(1H)-quinolones (AQ). The PqsR/PQS complex also generates transcription of *phnA* and *phnB* genes. These genes encode synthases involved in anthranilate production, the starting substrate of the HHQ/PQS metabolic route.

The TesB thioesterase can replace the metabolic functions of the PqsE enzyme. *P. aeruginosa* uses this pathway to produce over 50 other metabolites, including HQNO and DHQ secondary metabolites. These metabolites are toxic to prokaryotic and eukaryotic cells, as shown in Fig. (**29**) [127, 128].

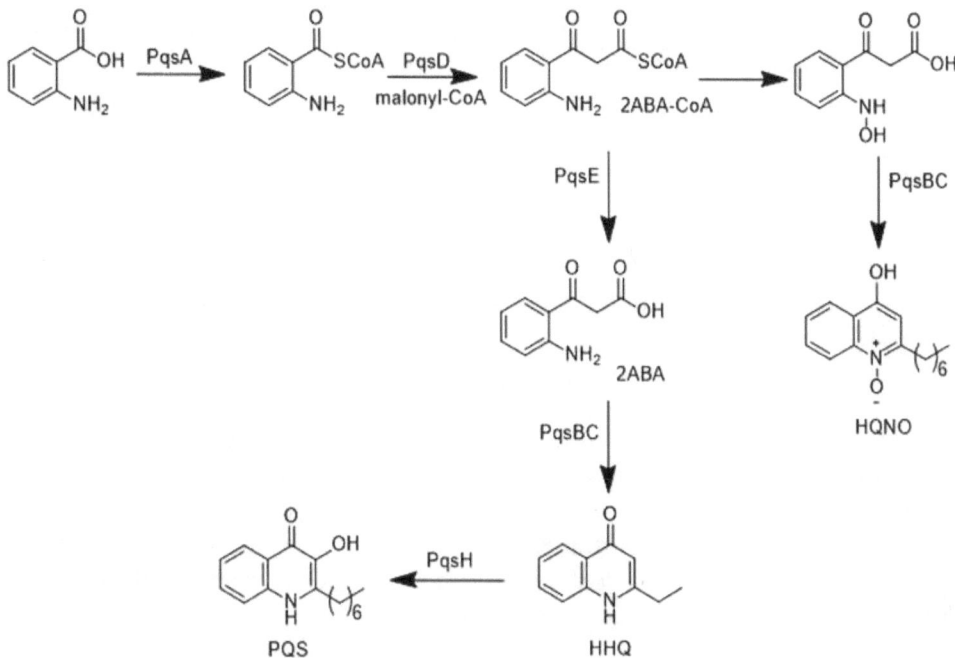

Fig. (29). Biosynthesis of 2-alkyl-4(1H) quinolone signaling molecules.

The *pqs* circuit stimulates pyocyanin secretion and various virulence factors in biofilm formation, such as rhamnolipids and exopolysaccharides (Pel and Psl). However, it also produces elastases A and B, outer membrane vesicles (OMV) transporting PQS, lectins A and B, hydrogen cyanide, and efflux pumps. Inducing rhlI expression upregulates the rhl circuit. The BHL/RhlR complex suppresses the expression of the *pqsR* gene, while the odDHL/ LasR transcription factor upregulates it, as does the *pqsH* factor [127].

It has been reported that the pqs QS system showed a prominent involvement in the virulence regulation of *P. aeruginosa*. Compared to the other QS systems in *P. aeruginosa*, the *pqs* system did not show any disadvantages associated with the *las* and the *rhl* systems. In chronic infections caused by *P. aeruginosa*, the *las* regulatory circuit is the first QS (quorum sensing) system to be lost. Regarding the *rhl* system, its effect on virulence modulation is observed to be non-unidirectional. RhlR agonists decrease the production of pyocyanin but increase the production of rhamnolipids. Conversely, RhlR antagonists have the opposite effect, increasing pyocyanin production and decreasing rhamnolipids.

In infection, *P. aeruginosa pqs* QS is crucial in regulating many virulence factors. Astonishingly, the expression of 182 genes is adjusted in reaction to exogenous PQS [129]. It has been suggested that PQS is affected by either directly acting through PqsR or PqsR-independent mechanisms. The latter is most likely due to PQS's ability to chelate iron and act as an antioxidant. Moreover, research has revealed that the thioesterase PqsE is a key effector molecule of pqs QS despite its dispensable biosynthetic role in *P. aeruginosa*. This is because alternative thioesterases exist [129].

PqsD, the second enzyme in the biosynthetic cascade of 2-heptyl-3-hydro-y-4-quinolone, has been studied intensely. Several design strategies have been pursued, leading to diverse structural class inhibitors. The first reported inhibitors of PqsD were 2-benzamidobenzoic acids.

The PqsE enzyme is responsible for regulating the activity of 145 genes that are crucial for pathogenicity traits. However, the exact mechanism by which it achieves this remains unknown. Interestingly, only 30 of these genes overlap with the PQS regulon, indicating that PqsE and PQS are essential in the *pqs* QS response. Together, they control the expression of genes responsible for various pathogenicity traits, such as the production of enzymes that lead to the biosynthesis of phenazine, hydrogen cyanide, and other factors involved in biofilm formation. Additionally, PqsE and PQS regulate the synthesis of enzymes that produce rhamnolipid, efflux pumps, secretion components, exotoxins, and siderophore synthases. Although PqsE is an attractive target for controlling *pqs* virulence, the mechanism behind its regulatory function remains unknown and requires further investigation. Therefore, it is necessary to elucidate the regulatory mechanism of PqsE to understand its role in pathogenicity and explore its potential as a target for controlling pqs virulence [128].

Multiple studies have shown that combining *rhl* and *pqs* targeting QSI can provide anti-virulence effects. Rhl and Pqs are known to cause virulence factor production in environments with limited phosphate and iron, while Las has a minor impact. Therefore, simultaneous inhibition of Rhl and Pqs can reduce virulence when Las inhibition is ineffective. These findings can predict QS inhibitor activity in critical infection situations, such as cystic fibrosis sputum. These outcomes showed that in *P. aeruginosa* and other pathogens, the environmental signals could impact the efficiency of small-molecule QS inhibitors. The *pqs* system is active in cystic fibrosis patients with chronic infections, and blocking this master regulator can lead to an unambiguous anti-virulence effect [130].

Various studies on the *pqs* system have focused on two molecular targets: the signal molecule synthase PqsD and the receptor PqsR (MvfR). Currently, there is more progress in projects targeting PqsR. It has been found that QSI (quorum sensing inhibitors) that target PqsR, a transcriptional regulator, have more potent anti-virulence effects than those that target PqsD, which is involved in the biosynthetic enzyme cascade. However, some research suggests that QSI, which targets PqsR and PqsD or PqsBC, can have a synergistic effect [131].

A new inter-cellular communication signal, called IQS, has been discovered that integrates environmental stress signals with the quorum sensing network. This signal molecule belongs to a new class of quorum-sensing molecules and has been identified as 2-(2-hydroxyphenyl)-thiazole-4-carbaldehyde. The genes responsible for IQS synthesis are in a non-ribosomal peptide synthase gene cluster (ambBCDE). When this gene cluster is disrupted, it leads to decreased production of PQS and BHL signals and reduces virulence factors, such as pyocyanin, rhamnolipids, and elastase. (Table **5**) [132].

When the QS system is activated, bacteria behave collectively. However, some act independently, such as the so-called cheating cells. For example, during the growth of *P. aeruginosa* PAO1 in the laboratory, LasR mutant cheaters emerged to avoid QS-regulated protease production. These mutant cells do not participate in protease production.

Table 5. QS systems, regulon and autoinducer of *Pseudomonas aeruginosa*.

	QS *P. aeruginosa*	Regulon	Autoinducer
LasR → IqsR; LasR → RhlR; LasR → PqsR; RhlR ⇌ PqsR; PqsR → IqsR	Las R	Las A protease Las B elastase Apr alkaline protease *Iqs* *Pqs* *Rhl*	N-(3-oxododecanoyl)-L-homoserine lactone (OdDHL)
	Rhl R	LasB elastase RhlAB rhamnolipids Phz pyocyanin Hcn hydrogen cyanide	N-butanoyl-l-homoserine lactone (BHL)
	PqsR	Phz pyocyanin Hcn hydrogen cyanide LecA lectin *Rhl*	2-heptyl-3-hydroxy-4-quinolone (PQS)
	IqsR	*Pqs* *Rhl*	2-(2-hydroxyphenyl)-thiazole-4-carbaldehyde (IQS)

LasR mutations are frequently found in *P. aeruginosa* isolates from the lungs of Cystic Fibrosis (CF) patients with chronic infections. However, unlike the PAO1 strain, many LasR mutants in chronic diseases have an active RhlIR quorum-sensing (QS) system. These mutants exhibit QS-regulated behaviors when cultured under laboratory conditions. The differences in QS regulatory networks between clinical isolates and PAO1 likely reflect the adaptations of *P. aeruginosa* during prolonged infections. These results indicate a possible change in *P. aeruginosa* QS hierarchy [133].

Various small regulatory RNAs (sRNAs) fine-tune bacteria like PhrD and RsmY responses to environmental signals and regulate quorum sensing. PhrD, unlike other sRNAs, regulates *P. aeruginosa* quorum sensing in diverse situations.

RsmY regulates the *rhl* system under nutrient-rich conditions. PhrS activates the PqsR quorum-sensing regulator under oxygen-limited conditions, increasing pyocyanin production. Under nitrogen limitation, the base pairing of NrsZ sRNA with *rhlA* activates rhamnolipid production. PhrD sRNA positively influences RhlR and enhances both rhamnolipid and pyocyanin production.

These findings demonstrate a significant pathway to promoting RhlR transcripts, which arise from various promoters before the PhrD interaction region. As a result, PhrD functions as an sRNA that helps the bacterium adjust to different environmental and host-induced stimuli [134].

-Quorum-sensing Regulation in *Staphylococci*

Staphylococcus aureus and *S. epidermidis* have been causative agents in various human infections, including skin and soft tissue, the bloodstream, the respiratory and skeletal systems, and diseases involving implanted medical devices—*Staphylococci*'s capacity to sense the bacterial cell density and quorum. One central system, the accessory gene regulator (Agr), controls the response to genetic adaptations.

The Agr extracellular signal is a modified peptide that contains a thiolactone structure. In conditions of high cell density, Agr increases the expression of several toxins and degradative exoenzymes while decreasing the expression of certain colonization factors. This regulation is essential at the time of expression of the virulence factor when infections occur and during the development of acute diseases. On the other hand, reduced Agr activity is associated with chronic staphylococcal infections involving biofilm formation. As a result, researchers are evaluating drugs that can inhibit Agr to control acute forms of *S. aureus* infections [135].

Escherichia coli

Escherichia coli is a bacterium that commonly inhabits the gut of humans and other animals. It uses QS to coordinate gene expression, which is involved in nutrient uptake and metabolism.

The *V. harveyi* system 1-autoinducer (AI-1) has been purified and identified as hydroxybutanoyl-L-homoserine lactone. Its synthesis depends on the genes *luxL* and *luxM*. However, a report about the structure or biosynthetic gene(s) for the system of two autoinducers (AI-2) must be provided. The sensor proteins that detect AI-1 and AI-2 are members of the bacterial family of two-component adaptive regulatory proteins, and the mechanism of signal transduction is a phosphorylation-dephosphorylation cascade. Unique strains of *V. harveyi* have been engineered to respond exclusively to AI-1 or AI-2. These strains are employed to create a bioassay that can identify autoinducers produced by other bacterial species. This bioassay has demonstrated that several *Escherichia coli* and *Salmonella typhimurium* strains produce an AI-2-like activity. When specific carbon sources are present, enteric bacteria produce the most AI-2 activity in the mid-exponential phase. Nevertheless, unlike further explained quorum-sensing systems, in *E. coli* and *S. typhimurium*, the AI-2 sign is degraded when the bacteria are in the stationary phase. In *E. coli* and *S. typhimurium*, several environmental alerts have been demonstrated to impact autoinducer production and degradation groups. Prompt logarithmic development, selected carbon bases, lower pH, and elevated osmolarity all improved autoinducer production. Whereas requirements of the stationary phase, the lack of a preferred carbon source, neutral pH, and low osmolarity induced degradation of the AI-2 signal. It has been found that protein synthesis is necessary to activate signal production and degradation in both *E. coli* and *S. typhimurium*. Interestingly, it was observed that the laboratory strain *E. coli* DH5a, commonly used in research, is incapable of producing or degrading AI-2. However, the gene responsible for AI-2 production and degradation has been identified in *V. harveyi, E. coli*, and *S. typhimurium* bacteria, and it is highly similar. This gene, luxS, defines a new family of autoinducer-producing proteins found in many bacterial species identified by genome sequencing projects [136].

-Gut Ecosystem and QS Molecules

The microbes in our gut react to QS molecules, communicate with our cells, and help maintain a healthy balance. In cases of inflammation caused by an imbalance of bacteria, it is essential to consider the impact of these molecules on our immune system. The most studied QS molecules are the type-1 auto-inducers of N-acyl-homoserine lactones (AHL). AHL molecules, initially discovered in pathogens like *P. aeruginosa*, have been detected in commensals and the intricate microbial communities in the mammalian intestinal tract [137].

Earlier studies indicated that *P. aeruginosa* 3-oxo-C12-HSL could affect the intestinal barrier by impairing the integrity of protein junctions, as observed on a Caco2 intestinal epithelial cell line. In the presence of the bacterial signaling molecule AHL, the permeability of the cell membrane to ions and large molecules significantly increases. This change alters the expression and location of proteins that play a crucial role in cell adhesion, such as occludin, E-cadherin, and zonula occludens-1. The MAPK signaling pathway mediates this alteration process, which necessitates phosphorylation of the junction proteins, ultimately disrupting the integrity of the cell junctions. It has been observed that the response to 3-ox--C12-HSL leads to a significant change in calcium signaling. Interestingly, C4-HSL does not affect the barrier integrity in the same way as 3-oxo-C12-HSL. Recently, a new type of AHL called 3-oxo-C12:2-HSL was discovered in the gut, which has been found to have anti-inflammatory properties on the Caco-2/TC7 cell line that is stimulated by interleukin-1β (IL-1β). This effect was demonstrated by a reduction in the secretion of IL-8 [134].

-Orphan AHL system.

The importance of the QS system is reflected, in certain circumstances, by cooperation between bacteria. Some LuxR-type receptors found in bacteria do not have a corresponding LuxI-type synthase in their genome. These receptors are orphan or "solo" LuxR-type receptors because they lack a self-produced, cognate AHL signal. Bacteria with these receptors are believed to use AHLs that other bacterial species produce in their environment. SdiA is a highly conserved orphan LuxR-type receptor. Enterohemorrhagic *Escherichia coli* (EHEC) bacteria have this receptor, which causes severe diarrheal diseases in humans. Cattle are the natural hosts of this EHEC. Approximately 75% of EHEC infections in humans come from beef and other bovine products. EHEC begins to colonize cattle when they first eat it. When EHEC reaches the rumen, SdiAEC is activated by AHLs, which other harmless bacteria produce. SdiAEC is active in the rumen and abomasum. After passing through the cow's digestive system, EHEC colonizes the recto-anal junction (RAJ), where there are no AHLs, and SdiAEC does not repress the LEE operon [138].

Gram-negative bacteria use N-acyl homoserine lactones (AHSLs) for communication, which is predominantly mediated by LuxR-type receptors. Recent studies uncovered aryl-HSLs, α-pyrones, and dialkylresorcinols as further chemical languages of Gram-negative bacteria. These findings extend the number of bacterial signaling molecules and suggest that cell-cell communication goes beyond acyl-HSL signaling.

Different languages and dialects are used in bacterial quorum sensing (QS) systems *via* distinct signaling molecules. Five classes of QS signaling molecules have been described until now. Modifications within the respective molecules represent the different dialects of one language (Table **6**).

Table 6. Different Classes of QS Signaling Molecules.

Compound Type	Synthase System	Receptor	Precursor
N-acyl homoserine lactones	LuxI	LuxR	SAM
Aryl homoserine lactone (aryl-HSL)	Rpal	RpaR	SAM
Photopyrones (PPYs) Dialkylresorcinols (DARs)	Ppys DarABC	PluR PauR	ACPs

(Table 6) cont.....

Compound Type	Synthase System	Receptor	Precursor
2-Heptyl-3-hydroxy-4-quinolone (PQS)	Pqs ABCD	PqsR	CoA anthranilic acid p-coumaroyl-CoA, thioester-

The corresponding synthase system and the cognate receptor of the five signaling molecules [139] are:

- LuxI/LuxR from *Vibrio fischeri,*
- RpaI/RpaR from *Rhodopseudomonas palustris,*
- PpyS/PluR from *Photorhabdus luminescens,*
- DarABC/PauR from *Photorhabdus asymbiotica,* and
- PqsABCD/ PqsR from *Pseudomonas aeruginosa.*

In *P. aeruginosa,* the LasI synthase produces the signal molecule N-oxododecanoyl-L-homoserine lactone (OdDHL or 3OC12-HSL), detected by the cytoplasmic receptor LasR. It is known that *P. aeruginosa* isolates, both in natural environments and clinical facilities, have a higher frequency of mutations in *lasR* than its cognate synthase gene, *lasI.* Approximately 40% defect in LasR. Of particular interest is that it is known that OdDHL increases pyocyanin production in the absence of LasR. It is possible that OdDHL does not exclusively serve as an autoinducer of LasR. Furthermore, despite the lack of LasR activity, the *rhl* system is still functional in a subset of LasR-defective strains, ensuring QS functionality and the expression of survival and virulence factors.

As previously thought, LasR may not be essential for cell-to-cell communication in P. aeruginosa. RhlR may play a more central role than LasR in regulating this process since LasR deficiencies are more common than RhlR deficiencies [140].

Quorum-Sensing Inhibitors

Many efforts have been made to regulate Quorum-Sensing (QS), either by decreasing the production of toxins or by activating QS at low cell density. This last regulation can help the immune system eliminate virulence factors. As a result, potential therapies for QS modulation have been developed, including macrolide antibiotics, QS vaccines, and competitive QS inhibitors [141].

Macrolides such as azithromycin inhibit QS by reducing the production of several virulence factors of *P. aeruginosa,* such as elastase, rhamnolipids, and alginate synthesis [142]. According to reports, using a lower dose of azithromycin did not affect the mRNA expression of specific genes related to quorum-sensing (*lasI, lasR, rhlI, rhlR,* and *vft, rsaL*). However, it did lower the expression of many enzymes involved in N-acyl homoserine lactone (AHL) production. Although the downregulation of these enzymes was only slight, it could still lead to a significant decrease in AHL production, ultimately inhibiting QS in *P. aeruginosa* [143].

It has been reported that azithromycin can interact with the outer membrane of *P. aeruginosa* PAO1. This interaction causes divalent cations to be displaced from their binding sites on LPS, which increases its permeability. This activity explains why macrolide antibiotics can effectively treat chronic airway infections caused by *P. aeruginosa* at sub-MIC levels. However, Mg^{2+} and Ca^{2+} can counteract this effect [144].

It has been proposed to use azithromycin as an adjunct therapy. However, it possesses positive and negative properties that cannot be overlooked. Indeed, there is developing evidence that the lung microbiome significantly influences the shaping of the host immune response and lung inflammation. It has recently been demonstrated that azithromycin improves bacterial metabolite presentation by modulating lung microbiota and metabolome. It is important not to underestimate the unfavorable effects on microbiota if the beneficial lung microbiome modulation occurs in the trial [145]. Azithromycin (AZI), chemically related antibiotics, erythromycin, and clarithromycin affect AHL production and reduce the production of virulence factors such as protease and elastase activity (Fig. **30**).

In biofilms, *Staphylococcus aureus* and *Pseudomonas aeruginosa* can cause chronic infections that are difficult to eliminate. *Pseudomonas aeruginosa*'s biofilm formation and virulence are controlled by pseudomonas quinolone-dependent QS. Targeting the transcriptional regulator PqsR (MvfR) with antagonists can reduce this development. A library of quinazolinone (QZN) compounds, including PqsR agonists and antagonists, was tested for its ability to combat *S. aureus* bacteria when co-cultured with *P. aeruginosa* and combined with the antibiotic tobramycin. One particular inhibitor, QZN34 (Fig. **31**), was found to kill Gram-positive bacteria but not Gram-negative ones. Additionally, QZN34 effectively precluded *S. aureus* biofilm construction and impaired *S. aureus* biofilms and disrupted the development of *P. aeruginosa* biofilms. When *P. aeruginosa* and *S. aureus* were in mixed biofilms, tobramycin did not work well against *S. aureus*. However, combining the aminoglycoside antibiotic and QZN 34 eliminated the mixed-species biofilm. It was found that QZN34 affects Gram-positive bacteria by disrupting their membranes and causing a loss of transmembrane potential [146].

Azithromycin　　　**Erythromycin**

Clarithromycin

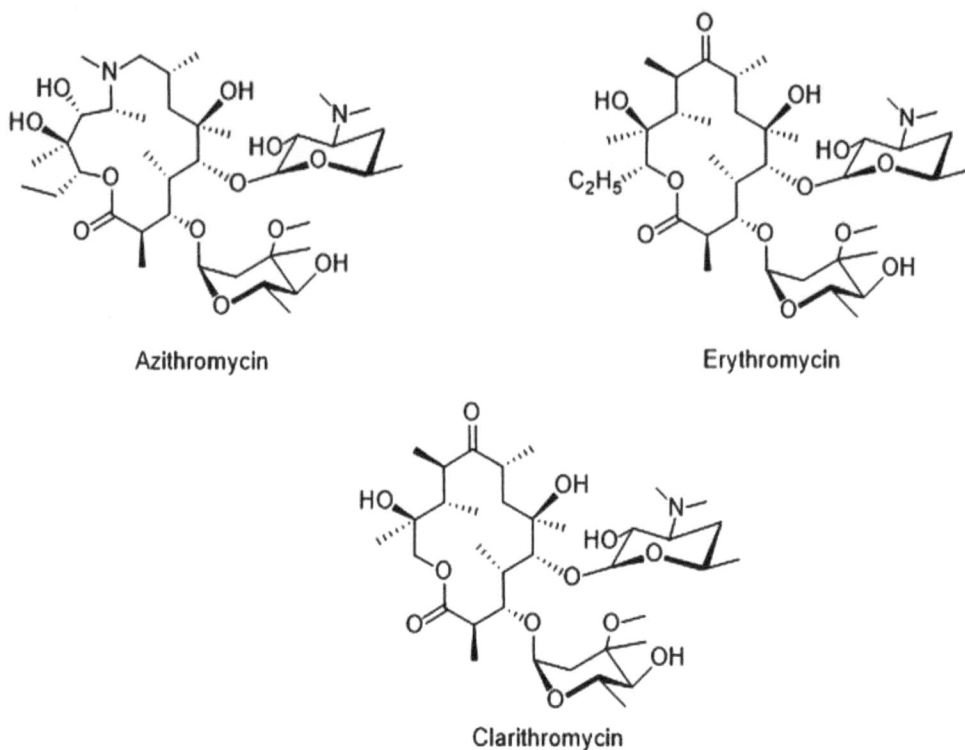

Fig. (30). Chemical structures of Azithromycin, Erythromycin A and Clarithromycin.

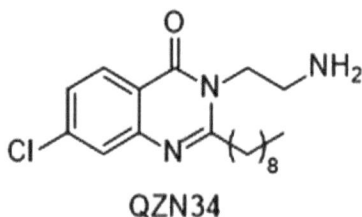

QZN34

Fig. (31). Chemical structure of QZN34.

Proof has been documented sustaining the utilization of some anticancer medications, including 5-fluorouracil (5-FU), gallium (Ga) hybrids, and mitomycin C as antibacterials. Individually, these medicines have some encouraging effects, such as comprehensive action (all three drugs), dual antibiotic and antivirulence effects (5-FU), effectiveness against multidrug-resistant strains (Ga), and the capacity to destroy metabolically static persister cells that generate chronic diseases (mitomycin C) [147].

Recently, a study screened a library of 1,600 drugs approved by the U.S. Food and Drug Administration and identified new inhibitors for the Pqs QS system in *Pseudomonas aeruginosa*.

Research has revealed that clotrimazole and miconazole, both antifungal medications, and clofoctol, a potent antibacterial compound that targets Gram-positive pathogens, can effectively inhibit the Pqs system. This activity is likely due to their ability to target the transcriptional regulator PqsR specifically. Clofoctol has been identified as the most effective inhibitor, significantly reducing the manifestation of *pqs*-controlled virulence characteristics in *P. aeruginosa*, including pyocyanin, swarming motility, biofilm formation, and genes associated with siderophore presentation. In addition, clofoctol has been established to protect *Galleria mellonella* larvae from *P. aeruginosa* disease and to inhibit the *pqs* QS system in *P. aeruginosa* strains found in cystic fibrosis patients. It is worth noting that clofoctol is already approved for treating pulmonary infections caused by Gram-positive bacterial pathogens, which makes it a promising candidate for treating *P. aeruginosa* lung infections as an antivirulence agent (**Table 7**) [148].

Pseudomonas aeruginosa has two genes, *lasI* and *rhlI*, that produce autoinducers (AI). LasI and RhlI synthesize N-(3-oxo-dodecanoyl)-homoserine lactone (OdDHL) and N-butyryl-l-homoserine lactone (BHL), respectively, which are their signals.

The *rhl* system plays a crucial role in regulating swarming motility, which is necessary for the early stages of biofilm establishment and the production of virulence factors such as rhamnolipid and pyocyanine. On the other hand, the *las* system controls genes related to biofilm development. The LasI and RhlI systems respond to autoinducers of the homoserine lactone type (AHL). It is evident from the above that the AHL plays a crucial role in developing the QS system (QSS). Consequently, removing the AHL will result in the cessation of the QSS [149, 150].

i. Reducing the activity of AHL synthase or AHL-related receptor protein.
ii. Inhibiting the production of QS signal molecules.
iii. Degradation of the AHL.
iv. Mimicking the signal molecules primarily by using synthetic compounds as analogs of signal molecules (Fig. **32**).

Table 7. The structure and stereo structure of PqsR inhibitors Clofoctol, Miconazole, and Clotrimazole.

Compound	Stereo structure*
 Clotrimazole antifungal	
 Miconazole antifungal	
 Clofoctol antibacterial	

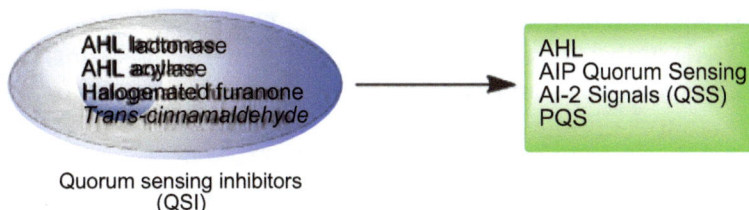

Fig. (32). Some Quorum-Sensing Inhibitors.

A previous study was conducted to identify potent AHL synthase inhibitors by testing both artificial and natural substances. It found three such inhibitors: salicylic acid, tannic acid, and trans-cinnamaldehyde (Fig. 33). After conducting tests with LC/MS, it was discovered that tannic acid and trans-cinnamaldehyde had a significant impact on suppressing AHL production by RhlI. Trans-cinnamaldehyde also decreased pyocyanin production in *P. aeruginosa* by up to 42.06%, inhibiting the Rhl QS system. Molecular docking analysis suggested that trans-cinnamaldehyde binds to LasI. The information gathered indicates that a new type of natural product may inhibit the activity of specific molecules involved in bacterial communication. This discovery could lead to developing a new method for fighting infections by disrupting bacterial communication [151].

Salicylic acid   *trans*-cinnamaldehyde

Fig. (33). Chemical structure of Salicylic acid and Trans-Cinnamaldehyde.

Furthermore, it has been reported that cinnamaldehyde effectively inhibits and disperses *P. aeruginosa* biofilm. When cinnamaldehyde combines with colistin, it shows a synergistic effect. On the contrary, combining cinnamaldehyde with carbenicillin, tobramycin (TOB), or erythromycin shows no synergistic effect. Cinnamaldehyde has decreased the expression of *lasB*, *rhlA*, and *pqsA* in GFP reporter assays. The highest concentration of cinnamaldehyde tested in all the QS reporter strains showed a reduction of approximately 70% in GFP production. In reporter assays, TOB also exhibited strong QSI when combined with cinnamaldehyde. When combined, cinnamaldehyde with colistin and cinnamaldehyde with TOB showed an additive effect in inhibiting biofilm (75.2% and 83.9%, respectively) and in dispersing preformed biofilm (about 90% for both) compared to individual treatments.

Therefore, combining cinnamaldehyde with colistin or cinnamaldehyde with tobramycin can be an effective alternative to individual drug therapies for mitigating *P. aeruginosa* infection [152].

The effect of *P. aeruginosa* (ATCC 10154) on biofilm formation, motility, AHL lactones synthesis, and Lasi/Lasr expression in the presence of carvacrol was investigated. Docking analysis determined interactions between carvacrol and LasI and LasR proteins. When exposed to carvacrol at a concentration of 1.9 mM, *P. aeruginosa* produced 60% fewer AHLs without any negative impact on cellular viability. This result suggests a decrease in LasI synthase activity. The detected AHLs, including C12, C6, and C4, are known to be linked to biofilm formation, motility, and pyocyanin production. Carvacrol was found to decrease the expression of *lasR* while leaving *lasI* gene unaffected. Additionally, computational docking revealed that carvacrol interacts with specific residues in the functionally active area of LasI and within the binding pocket of LasR of *P. aeruginosa*. These outcomes indicated that carvacrol reduced the virulence of *P. aeruginosa* by inhibiting LasI activity, with the concomitant reduction of the expression of LasR, biofilm, and swarming motility. This study shares valuable insights into how carvacrol can inhibit the virulence factors of *P. aeruginosa* by targeting quorum sensing at both the enzymatic and gene levels. These findings can help develop natural anti-QS products that impact pathogenesis [153].

-Degradation of AHL

Initial studies showed that *Variovorax* and *Bacillus soil* bacteria could enzymatically break down AHL, a QS signal. Since then, many other enzymes have been discovered to modify or break down AHL.

These enzymes represent three catalytic classes:
- The amidases (also referred to as amido-hydrolases or acylase) cleave acyl homoserine lactones (AHLs) at the amide bond and liberate fatty acid and homoserine lactone.
- The reductases that convert 3-oxo-substituted acyl-homoserine lactone to their cognate 3-hydroxyl-substituted AHL
- The lactonases that open the homoserine lactone ring.

They occur in bacteria, archaea, and eukaryotes (Fig. **34**). Some bacteria, such as *Pseudomonas*, can cleave their own AHL signal. Besides AHLs, other QS signals can be biologically degraded: 3-OH-PAME regulates virulence in Ralstonia, the DSF produced by Xanthomonas, and the PQS by *Pseudomonas*. Microbial activities can also degrade diketopiperazines non-specifically, *i.e.*, irrespective of their QS signal function [154].

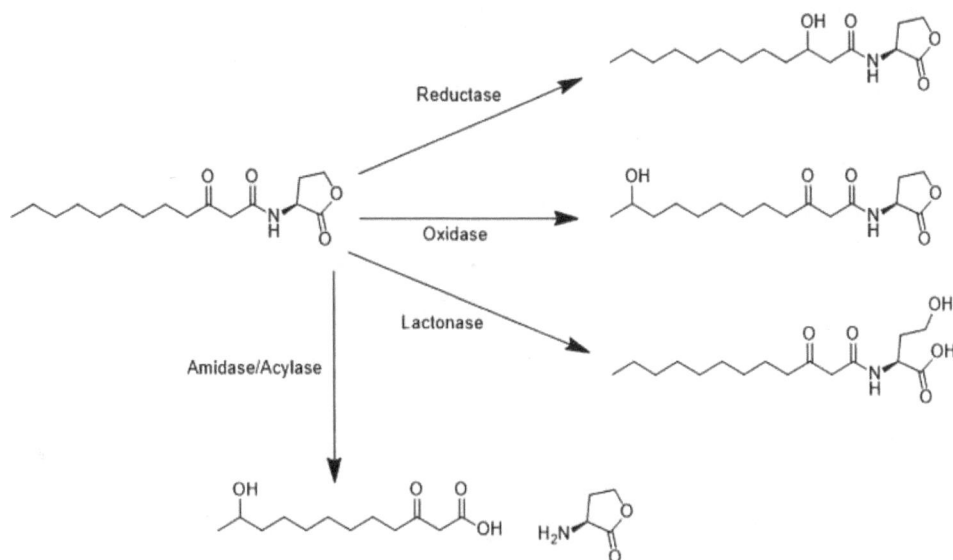

Fig. (34). Enzymatic transformation of AHL.

Pathogenic microbes regulate their virulence by producing signal molecules like acylated homoserine lactone (AHL) during quorum sensing (QS), which depends on cell density. However, some microbes produce enzymes called AHL-lactonases and—acylases that break down these QS signals, stopping the QSS.

Analyses of sequenced genome databases have shown organisms containing conserved domains for AHL-lactonases and –acylase: i) Streptomyces, ii) Deinococcus, iii) Hyphomonas, iv) Ralstonia, v) Photorhabdus, and a specific marine gamma proteobacterium. Genes for both enzymes within an organism were observed in *D. radiodurans* and others. These strains contain motifs for lactonase and acylase, supporting the observations. Gene sequences for AHL-lactonases and –acylase can be analyzed through phylogenetic analysis and multiple sequence alignment. Consensus sequences from this analysis can be used to design primers to amplify these genes, even in mixed cultures and metagenomes.

Quorum quenching (QQ) can prevent food spoilage, bacterial infections, and bioremediation [155]. The best target to disrupt the QS signals is AHL lactonase since it could be active against a wide range of AHLs with little influence on the length of the acyl chain and its efficiency [156].

-Synthesis of AHL Analogues

The possible ways that can be envisaged for developing the chemical analogs of the AHL molecule to act as QS inhibitors (QSIs) are:

a. Substitution(s) in the acyl side chain.
b. Alteration(s) in the lactone ring; or
c. Changes in both components (Fig. **35**).

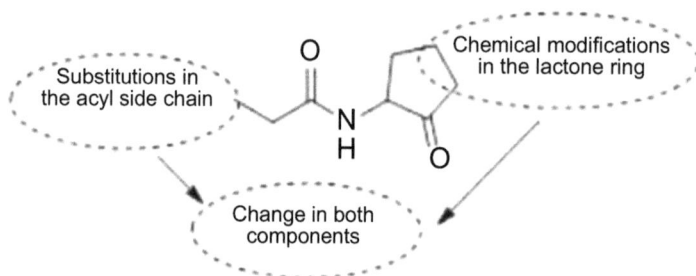

Fig. (35). Proposed Chemical Modifications of the AHL Nucleus.

Various derivatives with different substitutions in the side alkyl chain have been synthesized to achieve an analogous with greater power while preserving the furanone nucleus. In addition, some thiolactone derivatives have also been synthesized.

The use of racemic mixtures in the bioassay can be seen as a disadvantage because the stereocenter in HSL has already been identified as crucial for binding affinity. Enantiomerically pure sulfonamide derivatives were synthesized and tested in a LuxR monitor system (Fig. **36**). The study compared sulfonamide enantiomers and found that three were significantly more potent. Notably, both 7 and 8 exhibited comparable low activity levels compared to 6, implying that endocyclic oxygen in 6 impacts receptor recognition that is not seen with thiolactones. Except for Trp57 (in TraR), it is not clear which amino acid residues are critical for the spatial arrangement of the lactone moiety.

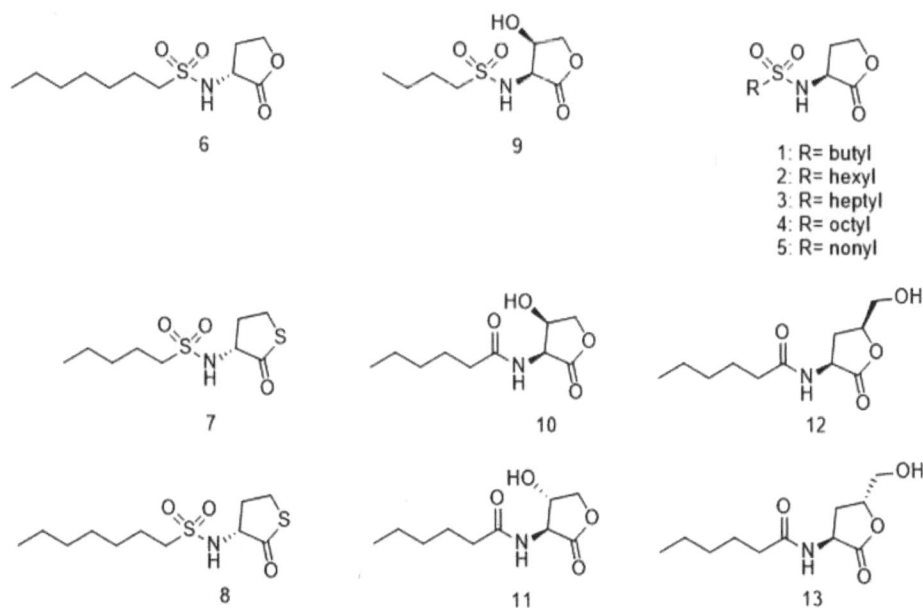

Fig. (36). Chemical structures of some QSI.

The results clearly state that a potency of 9 is just as effective as a potency of 1-5, with the only difference being the presence of a 4-hydroxy group. It is worth noting that analogs 10 and 11, which contain native carboxamide instead of sulfonamide, act as agonists of LuxR despite having the same acyl chain as LuxR antagonist 2.

The opposite effect was observed for 12 and 13, blocking the receptor in both cases. The 10 and 12 derivatives were more potent than their epimeric 11 and 13 derivatives (Fig. **36**). The reason for the effect of a 4-hydroxy and 5-hydroxymethyl group has yet to be established. However, the pattern can be interpreted as favorable interactions between the hydroxy group and an amino acid residue in the binding pocket (*e.g.*, Trp57 in TraR), normally interacting with the ester group of the lactone. Compound 15 (Fig. **37**), structurally related to 10 and 11, bearing the same acyl chain as the native agonist of LasR (8), attenuates the QS response in *P. aeruginosa*.

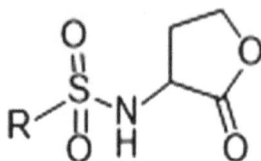

Compound	R
14	C_3H_7
15	C_5H_{11}
16	C_4H_9
17	C_6H_{13}
18	C_7H_{15}
19	C_8H_{17}
20	C_9H_{19}
21	$PhCH_2$
22	$Ph(CH_2)_2$
23	$Ph(CH_2)_3$

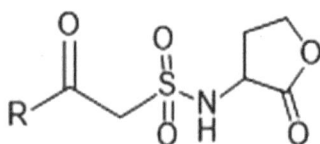

Compound	R
24	C_3H_7
25	$PhCH_2$

Fig. (37). Chemical structures of N-alkyl-, N-aryl. Sulfonamides are inhibitors of LuxR regulator.

Notably, Acyclic or cyclic alkyl substituents, especially those developed by replacing C3 with S in the acyl side chain, resulted in analogs that can block the expression of LasR-controlled QS receptors [157].

It has synthesized ten novel N-acyl-homoserine lactone variants by substituting the carboxamide bond with a sulfonamide moiety (Fig. **37**). The efficacy of these new analogs was assessed by their ability to inhibit the natural ligand (3-oxohexanoyl-L-homoserine lactone) of the LuxR transcriptional regulator, which governs the bioluminescence in *Vibrio fischeri* bacteria.

The evaluation of the 15 and 24 N-alkyl sulfonamides showed that replacing the carboxamide function in natural autoinducers with the sulfonamide resulted in good antagonist activity. The derivatives test results indicate that analog 22 displayed significant inhibitory activity while analog 25 was inactive. Thus, a sulfonamide function in alkyl-substituted derivatives resulted in a pronounced antagonist activity, while the combination of 3-oxo and sulfonamide functionalities gave no inhibitory effect.

To evaluate the influence of the alkyl chain length on the activity, the alkyl sulfonyl-HSLs 14–19 was tested. The number of carbon atoms in the alkyl chain affects the antagonist activity. The shortest analog, 14 with a butyl chain, was poorly active. The maximum activity was found for compound 15 with a pentyl chain. Increasing the chain length from five to nine carbon atoms (compounds 15, 16-20) resulted in a correlated decrease in opponent activity.

Molecular modeling indicates that the latter precludes a cascade of structural rearrangements required for assembling the functional LuxR dimer [158].

The active enantiomer of LuxR-regulated quorum sensing (QS) autoinducer is L-3-oxo-hexanoyl homoserine lactone (OHHL), while its D isomer is generally considered inactive. To better understand L-specificity and explore if it also applies to certain analogs in the acyl-homoserine lactone (AHL) family, it was synthesized OHHL and several AHL analogs in both racemic and enantiomerically pure D and L forms (Fig. **38**). The synthetic compounds were evaluated to induce or reduce bioluminescence in the LuxR-dependent QS system. The results showed that L-isomers are the only or the most active enantiomers. However, the D-isomer cannot be considered inactive, particularly for the natural ligand of LuxR (OHHL) and the remarkably comparable AHL agonist analog 2. According to molecular modeling, the D-isomer can show some activity when its lactone moiety can twist and interact with critical residues in the binding site, allowing the lactone carbonyl group and the amide function to come into play [159]. Inhibition of QS functions was reported by AHL analogs N-acyl cyclopentyl amines (Cn-CPA). C10-CPA was the most effective QSI against the Las and Rhl QS systems in *P. aeruginosa* [160].

Serratia marcescens is a harmful bacterium that can cause severe hospital infections. *S. marcescens* AS-1 strain produces two chemical types: N-hexanoyl homoserine lactone and N-(3-oxohexanoyl) homoserine lactone. These lactones are crucial in facilitating communication among bacteria and regulating essential functions such as producing prodigiosin, swarming motility, and forming a protective biofilm. *S. marcescens* has also synthesized a sequence of N-acyl cyclopentyl amides with acyl chain sizes ranging from 4 to 12 and evaluated their inhibitory effects on prodigiosin displayed in AS-1. A molecule known as N-nonanoyl-cyclopentyl amide (C9-CPA) has been found to strongly inhibit prodigiosin production.

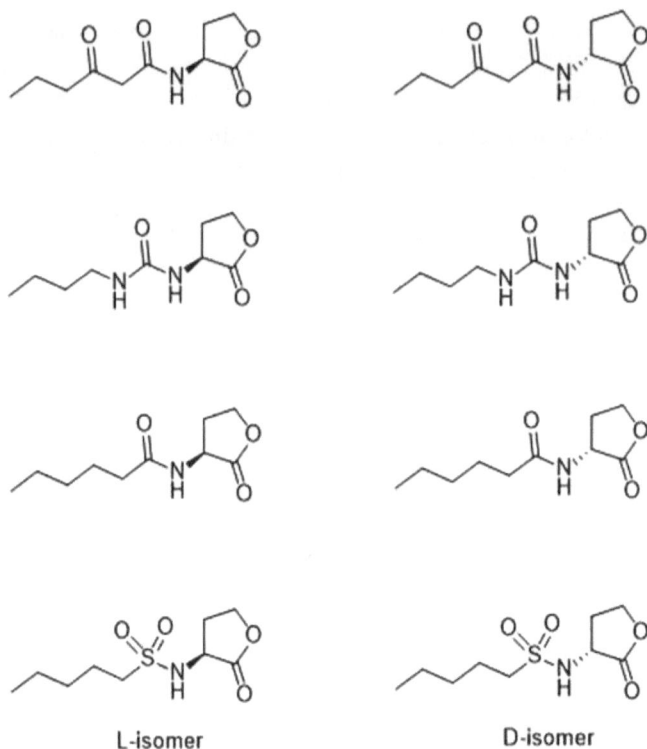

Fig. (38). Comparison of L and D AHL isomers.

Research has shown that C9-CPA can hinder *S. marcescens* AS-1's ability to swarm and form biofilms. When compared in a competition assay, it was found that C9-CPA is four times stronger than external C6-HSL when inhibiting quorum sensing. It is even more effective than the previously known halogenated furanone. Based on these findings, C9-CPA is a highly effective quorum-sensing inhibitor for *S. marcescens* AS-1 (Table **8**) [161].

It has synthesized a series of nine homoserine lactone-sulfonylurea derivatives with an alkyl chain residue, some bearing a phenyl group (Fig. **39**). AHL analogs bearing sulfonamide or a urea residue showed an inhibitory effect on *V. fischeri* QSS. It was primarily due to changes in tetrahedral geometries. N-pentyl-sulfonyl urea was the most active in terms of its inhibitory ability. In contrast, the phenyl group at the end of the alkyl chain was reported to be less active, suggesting that the phenyl-substituted compounds are less active than their aliphatic-substituted peers. Docking analyses performed on the most functional compound in a separate series of homoserine lactone-sulfonylurea derivatives indicate that the antagonist activity of these molecules could be connected to the alteration of the hydrogen bond in the ligand-protein complexes [162].

Table 8. Biological activities of some N-acyl-cycle pentyl amides (Cn-CPA).

	N-acylCPA	Bacteria	Activity
N-acyl-cyclopentyl amides	C10-CPA	*P.aeruginosa*	Effects against Las and RHl (QSS)
	C9-CPA	*S. marcescens* *Aeromonas hydrophila*	Inhibition of biofilm formation and motility,
	C6 CPA C7-CPA C8-CPA	*V. fischeri*	Reduction in luminescence

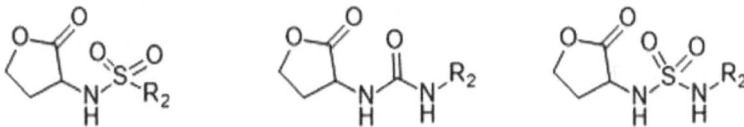

Fig. (39). Homoserine lactone sulfonylureas derivates.

A system for inhibiting quorum-sensing using recombinant bacteria has been developed and can be used for rapid screening. This system has been used to screen pure compounds and extracts from food and herbal medicine. During the study, it was discovered that there are multiple active QSI. The researchers focused on two of the most effective ones: garlic extracts (which contain at least three types of QSIs) and 4-nitropyridine-N-oxide (4-NPO). It was discovered that both garlic extract and 4-NPO have a concentration-dependent ability to inhibit QS. An analysis through GeneChip revealed that these substances significantly impact genes, particularly those related to virulence and regulated by QS. Both garlic extract and 4-NPO significantly lowered the pathogenicity of *P. aeruginosa* PAO1 in a *C. elegans* nematode model. In a clinical context, drugs based on QSI compounds might become interesting due to both the ability to downregulate the expression of virulence factors and the ability to make biofilms much more susceptible to conventional antibiotic treatment [163].

It is known that Quorum-sensing (QS) inhibitor 4-nitropyridine-N-oxide (4-NPO) decreases *Pseudomonas aeruginosa* biofilm formation. The 4-NPO is hydrophilic, with H-bond acceptors, non-H-bond donors, and is electrically neutral. All these attributes identified 4-NPO as an ideal antiadhesive compound. It was assumed that once 4-NPO is adsorbed to either the bacteria or solid surface, the physical-chemical properties of the solid surface and bacteria will be affected, decreasing the proportions of bacterial adhesion. The experimental determination of bacterial binding to silica using a microbalance of quartz crystal with dissipation (QCM-D) revealed that adsorption of 4-NPO to the silica exterior as well as to the external membrane of both *P. aeruginosa* PAO1(Gram-negative) and *S. aureus* (Gram-positive) 4-NPO decreases bacterial bonding to silica-coated surfaces. The proposed mechanism for reducing bacterial bonding is neutralizing the bacterial and silica cover charge by 4-NPO [164].

-Thiolactones.

Thiolactones and lactams are simpler versions of lactones, differing only by one atom in the backbone structure. Numerous research groups have discovered potent QS inhibitors that closely resemble AHL analogs and interact effectively with LuxR-type proteins. It has been reported that thiolactones and lactams, derived from native AHL signals, could restore the phenotype of *E. carotovora* involved in carbapenem biosynthesis at low concentrations. These findings demonstrate the effectiveness of these compounds as potential inhibitors of bacterial communication systems (Fig. **40**).

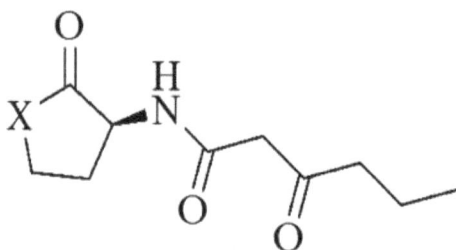

X= S agonist (10% activity relative to HSL)
X= NH agonist (0.15% activity relative to HSL)

Fig. (40). Thiolactone or lactam QS modulators in *E. carotovora*.

In a previous study, a selection of thiolactone analogs of AHLs was examined to determine their impact as QS agonists and antagonists of LuxR-type receptors (LuxR, LasR, TraR). It was shown that certain substances were effective against different bacteria, such as *V. ischeri, E. coli, P. aeruginosa*, and *A. tumefaciens* (Table **9**). Thiolactones A, B-F, particularly, have demonstrated potent antagonistic activity on the LuxR-type receptor, with IC_{50} values as low as the nanomolar scale. Some thiolactones have shown high activity, suggesting their potential as QSI [165].

Table 9. Thiolactone activity as QS agonists and antagonists of LuxR-type receptors.

	Compound	Activity
	A n=1	EC_{50}= 0.13 µM *(E. coli* LasR) EC_{50}= 13 µM (*P. aeruginosa* LasR) IC_{50}= 3.2 µM (TraR)
	B n=3	IC_{50}= 0.35 µM (Lux R) EC_{50}= 20 µM(TraR)
	C n=7	IC_{50}= 0.45 µM (Lux R) EC_{50}= 92 nM (*E. coli* Las R) EC_{50}= 32 µM (*P. aeruginosa* LasR) IC_{50}= 1.8 µM (TraR)
	D n=3	IC_{50}= 0.84 µM (Lux R) IC_{50}= 1,1 µM (*E. coli* Las R)
	E n= 4	IC_{50}= 0.31 µM (Lux R) IC_{50}= 0.79 µM (*E. coli* Las R) IC_{50}= 10 µM (TraR)
	F n=5	IC_{50}= 0.13 µM (Lux R) IC_{50}= 0.14 µM (*E. coli* Las R) IC_{50}= 2.8 µM (TraR)
	G n=9	EC_{50}= 1.9 µM (*E. coli* Las R) EC_{50}= 21 µM (*P. aeruginosa* LasR)

Pathogens often become resistant to antimicrobial agents through biofilm formation. The imidazole ring offers a promising scaffold for developing effective antimicrobial agents. A series of halogen-imidazoles was synthesized and evaluated against *P. aeruginosa*. In general, the antimicrobial activity increased as the position of the halogens varied from para to ortho derivatives. Among all the compounds, the bromo-derivatives showed moderate activity against most of the microbial strains in the study, as shown in Fig. (**41**). Chloro-derivatives are effective against bacterial and fungal infections. Among them, the orto chloride derivative has shown the most potent activity inhibiting the growth of Gram-negative and Gram-positive bacteria and fungi. In particular, the orto chloride compound has been shown to kill *P. aeruginosa*, even at low concentrations. Generally, fluorine derivatives were lousy in their antimicrobial action but were interestingly good against biofilms.

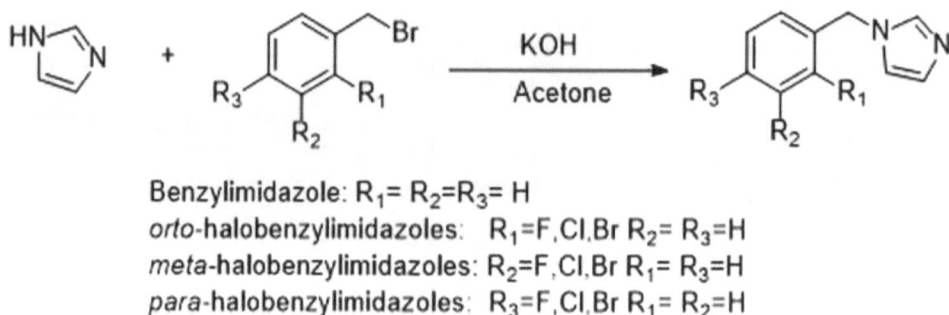

Benzylimidazole: $R_1 = R_2 = R_3 = H$
orto-halobenzylimidazoles: $R_1 = F, Cl, Br$ $R_2 = R_3 = H$
meta-halobenzylimidazoles: $R_2 = F, Cl, Br$ $R_1 = R_3 = H$
para-halobenzylimidazoles: $R_3 = F, Cl, Br$ $R_1 = R_2 = H$

Fig. (41). Structure of antibacterial halobenzylimidazoles.

Additionally, molecular docking showed that the compounds have good binding affinities for the LasR protein but are unlikely to be competitive inhibitors. This result suggests that the compounds may target proteins involved in biofilm formation beyond LasR, the primary transcriptional activator of quorum-sensing. Further tuning of the compounds is recommended to achieve improved activity [166].

-Halogenated furanones

Organisms naturally create secondary metabolites, including antibiotics, to survive. Microbes or fungi typically produce these natural substances, but can also be present in plants. It has been discovered that furanone-type secondary metabolites effectively deter other organisms' behavior. A similar hierarchical relationship between marine organisms and the red macroalga *Delisea pulchra* was proposed after discovering the secondary metabolites 26-29 (Fig. **42**). These structures represent 90% of the furanone content isolated and protected from the settlement of marine bacteria on its surface. More relevant here is that furanones 26-29 inhibited *S. liquefaciens* swarming due to the interference of QS.

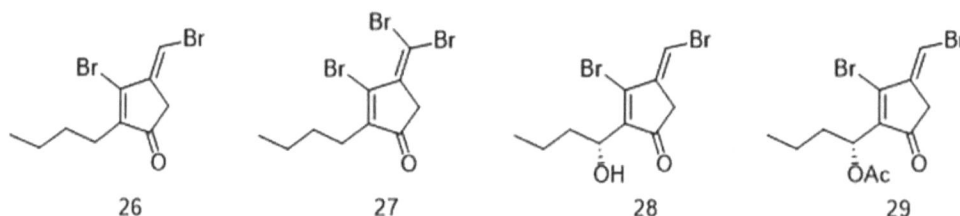

Fig. (42). Halogenated furanones isolated from red macroalga *Delisea pulchra*.

In the past, scientists believed that the structural similarity to AHLs was an essential factor in recognition by the transcriptional regulator. However, recent screening for LuxR and LasR antagonism has shown that the potency is not significantly affected by the butyl chain, and, importantly, it reduces *in vivo* activity. Such structures (*e.g.*, 30 and 31) have only the lactone scaffold in common with the AHLs. In addition, compound 26 has been found to have various effects beyond inhibiting Quorum Sensing in Gram-negative *P. aeruginosa* (although its activity in this area is weak). It has also been shown to inhibit growth in *V. anguillarum* and Gram-positive microbes like *Bacillus subtilis* and induce siderophore biosynthesis in *P. aeruginosa*. Studies have confirmed that halogenated ylidenebutenolides competitively inhibit QS, although other targets cannot be completely ruled out. Previous studies discovered that compounds 30 and 31 were more effective QS inhibitors than secondary metabolites 26-29, which influenced the direction of their subsequent research. The ability of 30 and 31 inhibited QS *in vivo* by quickly clearing *P. aeruginosa* infection in a pulmonary mouse model.

It has been observed that the effectiveness of halogenated ylidenebutenolides is reduced when they have an alkyl chain in the 3-position. This reduction is because the alkyl chain protrudes differently from the plane of the lactone ring compared to the acyl chain in AHLs. However, it is essential to note that other analogs of this type may exhibit different patterns.

According to the patent literature, 32, like 26, was more effective at inhibiting QS in a *P. aeruginosa* QS monitoring system than 33, which is like 30. Only 34 were as potent as 32, while the others were similar in potency to 21. It was discovered that the potency of these substances was not affected by the nitrogen substitution pattern. However, the unsaturation of the methylene group played a role in their activity, although the two most potent analogs (32 and 34) contained it (Fig. **43**) [157].

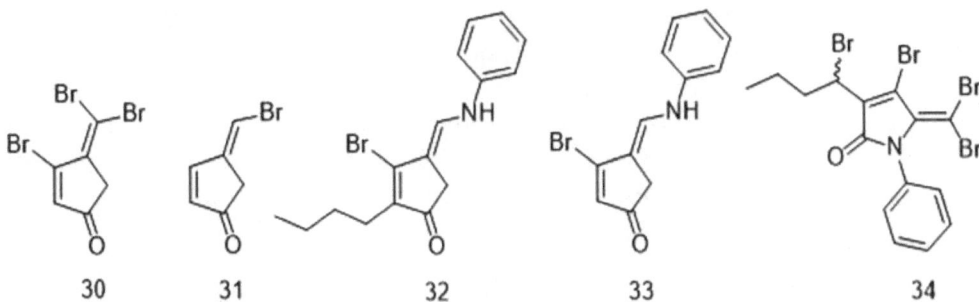

Fig. (43). Chemical structures of halogenated furanones 30-33 and pyrrolidone 34.

Previous studies have shown that the furanone C56 [(Z)-3-brom- -5-(bromomethylene) furan-2(5H)-one] explicitly blocks the expression of a PlasB-GFP reporter fusion while leaving growth and protein synthesis unaffected. It also decreases the production of essential virulence factors, suggesting a broad impact on the target genes in the final quorum-sensing pathway. Bacterial adhesion experiments to silica were performed. The furanone was spread to *P. aeruginosa* biofilms based on biofilm flow sections. The Gfp-based examination indicates that the compound infiltrates microcolonies and blocks cell signaling and QS in most biofilm cells. C56 has no impact on the initial attachment to the non-living surface. However, it alters the biofilm's structure and promotes bacterial detachment, which decreases the number of bacteria on the surface [167].

The benthic oceanic species *Delisea pulchra* yields halogenated furanones that can preclude the swarming movement of *S. liquefaciens* MG1. Adding these furanones from outside makes it possible to control the transcription of a gene regulated by quorum-sensing. It competes with the signal molecule N-butanoyl-L-homoserine lactone. This competition reduces the entire amount of serrawettin W2, a surface-active compound essential for the surface translocation of swarm cells. Furanones can also disrupt communication between species during swarming in mixed cultures [168].

AHLs are essential co-regulatory ligands that control the expression of genes responsible for virulence traits in different Gram-negative bacteria. New research has shown that AHLs can affect the levels of LuxR-type regulatory proteins within cells by binding to them and protecting them from being broken down by enzymes. This investigation attempted to find an *in vivo* relationship between a tritiated halogenated furanone and the LuxR enzyme, which showed a high production in *E. coli*. However, a permanent relationship between the algal metabolite and the LuxR enzyme was not located; the Western analysis revealed that halogenated furanones decreased the protein's half-life by up to 100 times. Based on the study, it has been found that halogenated furanones affect LuxR's action by destabilizing the AHL-dependent transcriptional activator instead of shielding it. The furanone-dependent declination in the cellular attention of the LuxR enzyme is associated with a reduction in the manifestation of a plasmid-encoded P(luxI)-gfp(ASV) fusion. This reduction indicates that the decrease in LuxR concentration is how furanones regulate the expression of AHL-dependent phenotypes [169].

It has been reported that the synthesis of a collection of 25 compounds consisting of 1'-unsubstituted and 1'-bromo or 1'-acetoxy 3-alkyl-5-methylene-2-5H)-furanones, along with two 3-alkyl maleic anhydrides (Fig. **44**). The abilities of these compounds were tested to prevent biofilm formation by *Salmonella Typhimurium* and regulate the bioluminescence of *Vibrio harveyi* through quorum sensing.

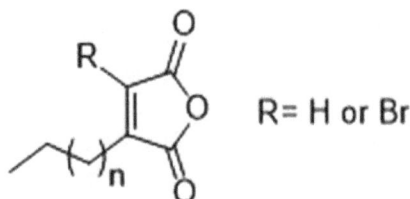

R= H or Br

Alkylmaleic anhydrides

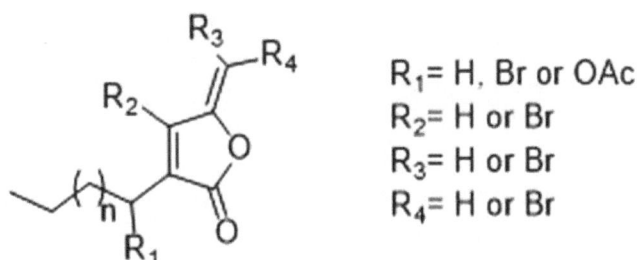

R_1= H, Br or OAc
R_2= H or Br
R_3= H or Br
R_4= H or Br

Brominated 3-alkyl-5-methylene-2(5*H*)-furanones

Fig. (44). Brominated furanones and alkylimaleic anhydrides.

The biological action of the 1'-unsubstituted furanones is particularly impacted by the size of the 3-alkyl chain and the bromination design of the ring frame. A Br atom on the 1' site of the 3-alkyl group significantly enhances the furanones' activity in both biological examination systems. The opening of an acetone function in this position does, in general, not improve the action. Finally, the potential of the (bromo) alkyl maleic anhydrides as an unexplored and chemically readily accessible class of biofilm and quorum-sensing inhibitors was revealed [170].

Furanones with varied side chain lengths and substitution in the ring structure, particularly those compounds that lacked a side chain but had an electronegative substituent on the furanone ring, effectively inhibited the QSS of *P. aeruginosa*. A previous study found that two furanones, namely (Z)-5-(bromomethylene) furan-2(5H)-one (C30) and (Z)-3-bromo-5-(bromomethylene) furan-2(5H)-one (C56), were highly effective in preventing surface colonization by *S. epidermidis*. These compounds have proven safe and could serve as valuable therapeutic agents [171].

Synthetic furanones have replaced the secondary metabolites in *Delisea pulchra*. These halogenated furanones boast powerful antibacterial properties and have been conclusively shown to inhibit AHL activity. Recent research has reported that synthetic halogenated furanone C56 reduces the expression of lasB, which is regulated by lasR, in a reporter gene assay. Additionally, it has been observed to positively impact extracellular elastase and chitinase activities on *P. aeruginosa* QS. Moreover, C56 could interfere with the QS of *P. aeruginosa* in the biofilm matrix and preclude biofilm maturation, while natural furanones showed limited activities. The effectiveness of compounds C56 and C30 in treating *P. aeruginosa* lung infections was evaluated in live mice models. The study found that both compounds improved bacterial clearance from the lungs, reduced colonization, lessened the severity of the disease, and significantly increased the survival time of the mice. A proposed mechanism of the action of C56 and C30 is the interaction between these compounds with the *rhl* system. It has been demonstrated that compound C30 has no activity on the LasR receptor. It has been reported that new synthetic furanones have potential benefits. One of these compounds, C-DB [3,4-dibromo-5-oxo-2,5-dihydrofuran-2-yl 2-(3,- -dimethoxyphenyl) acetate], has been shown to inhibit biofilm formation and reduce pyocyanin production by interfering with LasR. Compound C-DB also has low toxicity and good properties for further development (Fig. **45**) [128].

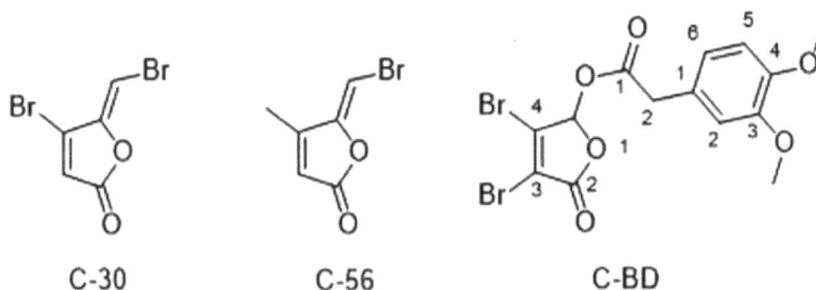

Fig. (45). Chemical structures of the halogenated furanones C-30, C-56, and C-DB.

An alternative approach to fighting antibiotic resistance is using quorum-sensing inhibitors (QSI). These compounds reduce the virulence of pathogens without killing them, and it is believed that they are less likely to lead to resistance. However, more experimental data are needed to support this hypothesis. It is also important to note that such studies are often conducted under conditions that may not accurately reflect real-life situations. For example, research has shown that the use of furanone C-30 on biofilms of *P. aeruginosa* increases their susceptibility to tobramycin compared to untreated biofilms. Additionally, studies have found that furanone C-30 can significantly reduce mortality rates in rainbow trout infected with *Vibrio anguillarum*. Furthermore, *Serratia liquefaciens* has been found to produce halogenated furanones to inhibit bacterial QS and prevent surface colonization.

The efficacy of integrating QSI furanone C-30 with the aminoglycoside antibiotic tobramycin in eliminating *P. aeruginosa* biofilms developed in an artificial cystic fibrosis sputum medium without causing any evolutionary resistance was estimated. The results demonstrate that the tobramycin/furanone C-30 mix reduced effectiveness behind five therapy cycles. The tobramycin/furanone C-30 mix induced the antimicrobial vulnerability of *P. aeruginosa* to tobramycin after 16 cycles of therapy, decreasing 8-fold.

Additionally, microcalorimetry revealed that exposure to furanone C-30, tobramycin, or a mix of both impacted the metabolic activity of *P. aeruginosa*. The analysis of the strains that underwent the combination treatment revealed mutations in *mexT*, *fusA1*, and *parS*. These genes are linked to antibiotic resistance. Moreover, when *P. aeruginosa* was solely treated with furanone C-30, a deletion in the *mexT* gene was also detected. These data suggest that furanone C-30 is not "evolution-proof" and quickly becomes useless as a tobramycin potentiator [172].

A valuable approach for combating bacteria resistant to multiple drugs is targeting their quorum sensing and type III secretion systems. These are both highly desirable targets for anti-virulence strategies. While halogenated furanones are known to reduce virulence associated with quorum sensing, it still needs to be made clear how effective they are at inhibiting the T3SS. The synthetic (Z)—4-bromo-5- (bromomethylene) -2 (5H) (C-30) and 5- (dibromomethylene) -2 (5H) (named hereafter GBr) were evaluated to preclude the secretion of type III exoenzymes, and QS-controlled virulence factors in *P. aeruginosa* PA14 and two clinical isolates (Fig. **46**). In addition, using a mouse model of skin abscess, it was found that GBr furanone effectively prevented bacterial growth. Furthermore, GBr was more practical, if not more effective, than C-30 against virulence factors.

In addition, GBr inhibited the T3SS, precluding the emission of effectors such as ExoS and others in the total of evaluated strains. When GBr at 50 and 25 µM doses was dispensed to CD1 mice infested with the PA14 strain, necrosis in the inoculation area appreciably decreased. The bacteria's systemic prevention was more effective with GBr than C-30 at 50 µM doses. Docking analysis indicated that the active site of the QS LasR regulator enhances its affinity for the geminal position of bromine in GBr.

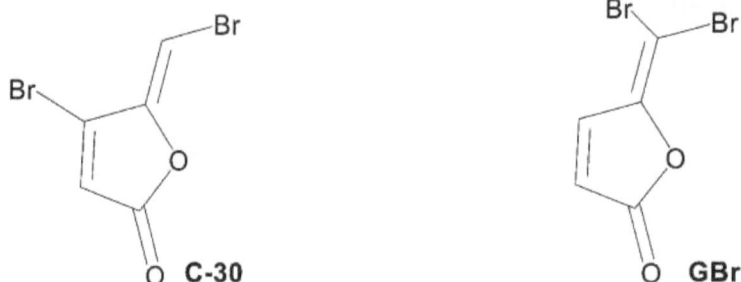

Fig. (46). Chemical structure of C-30 and GBr.

This research indicates that GBr furanone has multi-target properties that could lead to better anti-virulence treatments [173].

A previous study determined the connection between the ability to form biofilms, the expression of some critical virulence factors, and resistance to multiple drugs in *P. aeruginosa*. A total of 302 isolates were analyzed. To test phenotypic classification, the isolates underwent antimicrobial exposure. The outcomes cataloged the isolates into distinctive resistance types and various antibiotic resistance. A crystal violet microtiter plate-based method determined the isolates' ability to produce biofilm. The inhibition of virulence factors was also measured.

Pearson correlation coefficients (r) were used to calculate virulence factors. The results showed that ceftazidime, levofloxacin, and ciprofloxacin had the highest resistance rates (Table **10**).

Table 10. Resistance rates of ceftazidime, levofloxacin, ciprofloxacin and colistin.

Antibiotic	% Resistance Rates	Number of Isolates
ceftazidime	56.95	172
levofloxacin	54.97	166
ciprofloxacin	54.64	159
colistin	1.66	5

On the other hand, 133 isolates (44.04%) were classified as multidrug-resistant (MDR). Meanwhile, 60 (19.87%), 63 (20.86%), and 179 (59.27%) isolates were cataloged as weak, moderate, and robust biofilm creators, respectively. Interestingly, excluding pyocyanin display, the comparison between MDR and non-MDR isolates did not reveal substantial contrasts in the expression of virulence factors. Furthermore, no relevant correlations among the rate of biofilm development, pigment fabrication, or motility were detected.

Information on the interaction involving the existence and mechanisms of drug resistance, biofilm construction, and virulence is essential for managing chronic bacterial diseases and supplying methods for their control [174].

"Quorum quenching" (QQ) was created to define all processes that interfere with QS. A conclusive relation between QQ and its exploitation for competitive advantage has yet to be established. So far, the approach of QS inhibition has met with limited success [157].

Although QQ could inhibit some virulence factors and may be effective in altering the biofilm structure and increasing its susceptibility to antibiotic action [175], caution must be exercised in this strategy's practical application. *P. aeruginosa* has many ways to cause harm, such as creating biofilm and producing pigments, exotoxins, proteases, flagella, and secretion systems.

QQ is a harmless method that does not put much pressure on bacteria's survival. Another great feature of this method is its adaptability. This procedure has multiple molecular targets, including signal synthase, signals, and signal sensors. Furthermore, the best part is that QQ is broader than just regulating QS through AHL-based methods, as recent studies have shown.

Lastly, it is worth noting that the process involves using "environmentally friendly" components. The AHL degradation enzymes are naturally found in bacteria commonly present in soil or plant environments, and most of them are nonpathogenic isolates. The sensor antagonists are natural compounds from various plants, including edible plants. The AHL analogs used to select strains or consortia that degrade AHL are nontoxic compounds and are most often rapidly degraded by these bacteria. These features are crucial for the ever more demanding registration procedures of chemical compounds, such as the REACH directive. Aside from what has been previously mentioned, the biological importance of the QQ phenomenon is still fascinating. Although clues are often limited, it is believed that mammalian cells with the ability to degrade AHL are typically in contact with microbes. Nonetheless, a biological function for QQ has been identified in various species of the *Bacillus* and *Agrobacterium genus*, both Gram-positive and Gram-negative. However, these remain exceptions rather than the rule.

Nevertheless, the widespread quenching ability of various prokaryotic or eukaryotic microbes and that of plants strongly suggests that the quenching phenomenon must be reintroduced in the global QS regulation scheme. One apparent reason to do so is related to the very nature of a biological signal. To be perceived as such, a molecule must appear and disappear, either by being unstable or undergoing some turnover. Therefore, the QS regulation scheme must include, between synthesis and sensing, the various quenching processes that will likely limit the accumulation of the signals or their sensing and favor the obligatory degradation of these signals in nature. In other words, QS and QQ are opposite but complementary processes [176].

In conclusion, it has been proposed, based on experimental evidence, that the combination of antibiotics and QSIs may be the most effective treatment regimen [163].

NEW HYBRIDS AS ADJUVANTS AGAINST GRAM-NEGATIVE PATHOGENS

Despite the widespread awareness of the threat posed by bacterial strains resistant to multiple drugs, the FDA approved only two New Molecular Entities (NME) between 2010 and 2017, namely Dalbavancin and Orotavancin (both lipoglycopeptide-type drugs), that showed some effectiveness against drug-resistant bacteria (Table **11**). It is worth noting that in 2017, the WHO released ESKAPE, a list of the most critical drug-resistant bacteria [178].

Table 11. FDA-approved new molecular entities (NME) antibiotics from 2010 to November 2017.

Year	New Molecular Entity (NME) Antibiotic(s)	Class(es)	Route of Drug Administration
2010	Ceftaroline fosamil*	Cephalosporin	Systemic
2011	Fidaxomicin**	Macrolide	No systemic
2014	Dalbavancin**	Lipoglycopeptide	Systemic
2014	Oritavancin**	Lipoglycopeptide	Systemic
2014	Tedizolid phosphate*	Oxazolidinone	Systemic
2014	Ceftolozane-tazobactam*	Cephalosporin + β-lactamase inhibitor	Systemic
2014	Finafloxacin otic suspension*	Fluoroquinolone	No systemic
2015	Ceftazidime-avibactam*	Cephalosporin + β-lactamase inhibitor	systemic
2017	Delafloxacin*	Fluoroquinolone	Systemic
2017	Meropenem-vaborbactam*	Carbapenem + β-lactamase inhibitor	Systemic
2017	Secnidazole*	Nitroimidazole	Systemic

Capacity to effectively treat bacterial infections caused by antibiotic-resistant Gram-negative ESKAPE bacteria: * yes, **no.

Between 2018 and 2019, the US FDA approved several antibacterial agents after ESKAPE. These included plazomicin, eravacycline, sarecycline, omadacycline, and rifamycin 2018. In 2019, the combination of imipenem, cilastatin, and relebactam, along with pretomanid, lefamulin, and cefiderocol, was approved.

Plazomicin is an aminoglycoside antibiotic that targets Enterobacteriaceae infections, primarily for complicated urinary tract infections. Eravacycline is a fully synthetic fluorocycline that is approved for complicated intra-abdominal infections. Sarecycline is a tetracycline-derived antibiotic that might be useful for managing non-nodular moderate to severe acne. On the other hand, omadacycline, another tetracycline-derived antibiotic, may be utilized for patients with sensitive bacterial skin, skin structure conditions, and community-acquired pneumonia. Rifamycin, an RNA-synthesis suppressor, is supported for noninvasive E. Coli-caused travelers' diarrhea.

Two hybrid procedures were authorized for complex urinary tract disorders, intricate intra-abdominal infections (imipenem, cilastatin, and relebactam), and lung tuberculosis (combinations of pretomanid, bedaquiline, and linezolid). The semisynthetic lefamulin is a pleuromutilin antibiotic for community-acquired bacterial pneumonia. Cefiderocol, a cephalosporin antibiotic, is the latest antibacterial drug authorized in 2019 for complex urinary tract infections. Despite these new developments, there is still a pressing need to develop novel antibiotic strategies and drugs to overcome bacterial antibiotic resistance [179].

Pseudomonas aeruginosa, a Gram-negative bacterium, is a critical "priority 1" pathogen. This bacterium produces biofilm, which increases its resistance to drugs by up to 1000-fold. Infections caused by *P. aeruginosa* in the urinary and respiratory tracts can lead to severe complications in immunocompromised individuals. Furthermore, *P. aeruginosa* infections are frequently found in ICU patients affected by COVID-19.

To treat *P. aeruginosa* contagions, aminoglycosides, polymyxins, β-lactam/--lactamase inhibitor (BL/BLI) combinations, and fluoroquinolones (FQs) are typically used. β-lactams (BLs) and FQs are the most commonly prescribed antibiotics to treat a broad range of Gram-positive and Gram-negative bacteria. β-lactams bind to penicillin-binding proteins (PBPs), compromising the bacterial cell wall's integrity and reducing the organism's proliferation ability. However, resistance mechanisms towards BLs include modifications to PBPs' active site, low outer membrane (OM) permeability, and the production of β-lactamases that hydrolyze the four-membered ring of BLs.

Bacteria are classified into Gram-positive and Gram-negative types based on cell membrane structure. Gram-positive bacteria have a thick cell wall consisting of layers of peptidoglycan and teichoic acid anchored to the cytoplasmic membrane. In contrast, the structure of Gram-negative bacteria presents a light peptidoglycan coating enveloped by an internal and external membrane, forming the periplasmic space. This double-layered protection, abundance of efflux pumps, and highly selective porins pose a significant challenge to antibacterial agents targeting Gram-negative bacteria intracellularly.

Fluoroquinolones (FQs) are antibacterial agents that exert their antibacterial activity by inhibiting two bacterial enzymes, DNA gyrase and DNA topoisomerase IV. However, Gram-negative bacteria can develop multiple resistance mechanisms to FQs, which include mutations in the active site of DNA gyrase and DNA topoisomerase IV, overexpression of efflux pumps, and suppression of porin expression. These resistance mechanisms pose a significant challenge for antipseudomonal agents, as *P. aeruginosa* has extremely low OM permeability compared to other Gram-negative organisms.

The outer membrane in Gram-negative bacteria restricts molecular passage efficiently. An asymmetric bilayer, made up of two leaflets, is present in Gram-negative bacteria. The inner booklet comprises phospholipids, while the outer leaflet contains abundant lipopolysaccharides (LPS). The LPS molecule has three domains: lipid A, core oligosaccharides, and O antigen. The hydrophobic lipid A forms a lipid bilayer with the inner leaflet, while the hydrophilic body oligosaccharides and O-antigens open outwards to the extracellular territory. These territories are reliable for cellular identification and virulence, among other functions. Gram-negative bacteria's outer membrane makes them inherently resistant to several antibiotics, especially those with high molecular weight and hydrophobicity.

A novel approach to combat antibiotic resistance is using outer membrane permeabilizers (OMPs). These OMPs help to break down the outer membrane of Gram-negative bacteria, making it easier for antibiotics to enter. The membrane comprises phospholipids on the inner leaflet and lipopolysaccharides (LPS) on the outer leaflet, which act as a barrier to antibiotic penetration. OMPs destabilize the LPS by interacting with lipid A's negatively charged phosphate groups and removing divalent cations that link adjacent LPS molecules. Polybasic molecules like aminoglycosides and polymyxin derivatives are commonly used as OMPs to fight against Gram-negative bacteria.

Tobramycin (TOB) (Fig. **47**) is an aminoglycoside with two different action modes against *Pseudomonas aeruginosa*, depending on its concentration. At concentrations equal to or greater than 8 µg/mL, TOB has a bactericidal effect on *P. aeruginosa* by destabilizing the lipopolysaccharide (LPS) membrane. At lower concentrations equal to or less than 4 µg/mL, TOB interacts with the 16S ribosomal RNA, which inhibits protein synthesis.

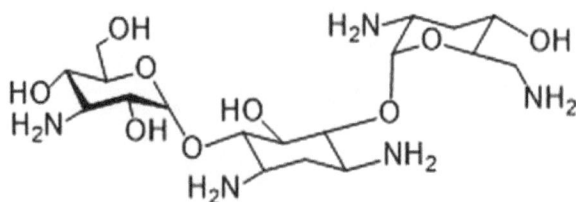

Fig. (47). Tobramycin chemical structure.

Extensive research has demonstrated that TOB-based outer membrane proteins (OMPs) can be synthesized through the linkage of TOB to various antibiotics, such as ciprofloxacin (CIP) (Fig. **48**), rifampicin (RIF), polymyxin, efflux pump inhibitor (EPI) 1-(1-naphthyl methyl)-piperazine (NMP), and ion-chelator cyclam (CYC), employing long carbon-chain linkers ranging from C8 to C12. These hybrid adjuvants exhibit an unparalleled synergistic effect with different classes of antibiotics, including fluoroquinolones (FQs), beta-lactams (BLs), and tetracyclines, against Gram-negative bacteria. Notably, these hybrid adjuvants have shown remarkable efficacy against *P. aeruginosa.*

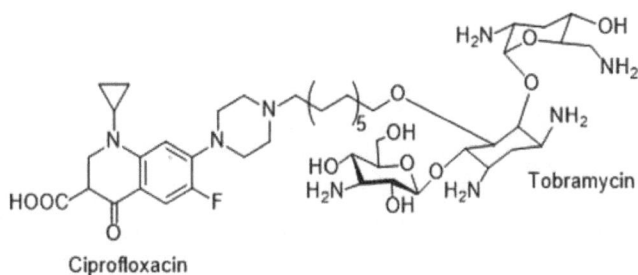

Fig. (48). Chemical structure of ciprofloxacin-tobramycin hybrid.

Studies have shown that these molecules disrupt the outer membrane through self-promoted uptake and reduce efflux pump activity by interfering with the proton motive force in the cytoplasmic membrane of *P. aeruginosa* [178].

To obtain hybrid adjuvants with better activity, the synthesis and evaluation of hybrids formed by three molecules, including tobramycin, ciprofloxacin, levofloxacin, and moxifloxacin, as TOB-CIP-NMP and TOB-TOB-CIP hybrids (Fig. **49**), have been reported.

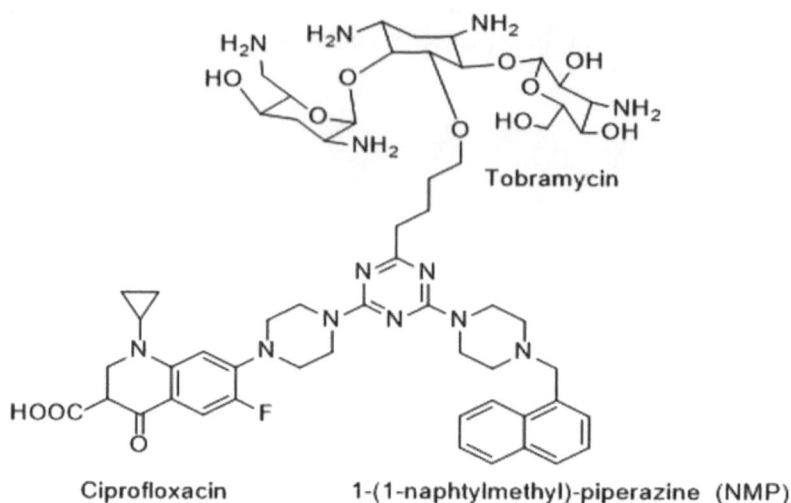

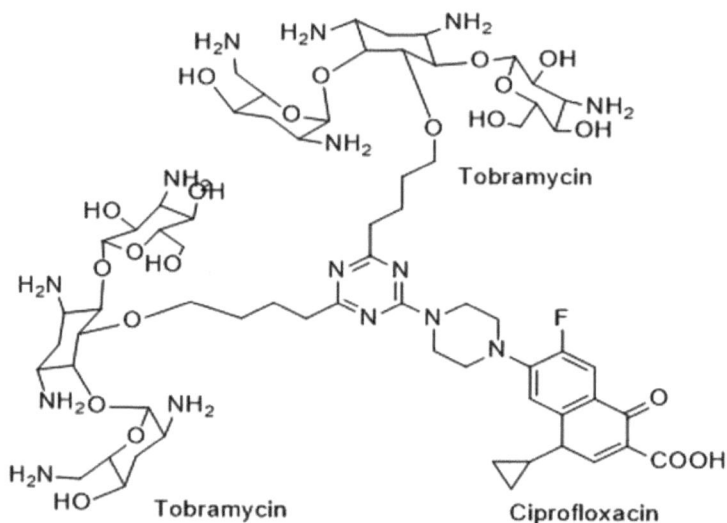

Fig. (49). Chemical structures of TOB-CIP-NMP and TOB-TOB-CIP hybrids.

The derivatives of TOB-based adjuvants belong to a new class of OMPs that significantly improve the effectiveness of multiple antibiotic classes. Previous studies have identified TOB-TOB-CIP as one of the most potent derivatives, synergizing with 16 out of 20 tested antibiotics. This synergizing includes Gram-positive-selective antibiotics like novobiocin, RIF, and niclosamide, as well as FQs and BLser, typically used for treating MDR GNB. The TOB-TOB-CIP hybrid effectively eradicates the MDR *P. aeruginosa* strain PA86052. These findings suggest that a triple combination, including derivative 4, may enhance the antibiotic effect of ATM/AVI against *P. aeruginosa*.

It has been observed that polymyxin B nano peptide (PMBN), which is considered the gold standard potentiator at a concentration of 8 µg/mL (7 µM), or the TOB-TOB-CYP hybrid at a concentration of 4 µM, exhibited comparable potentiation of several antibiotics, including novobiocin, rifampicin, chloramphenicol, linezolid, clindamycin, FQs, and BLs, in PAO1. However, PMBN was more potent than other TOB-hybrids when used with antibiotics like vancomycin, trimethoprim, pleuromutilin, erythromycin, and tetracyclines. Previous studies have shown that TOB alone did not exhibit synergistic effects with the same panel of antibiotics against PAO1. Additionally, the parent compounds NMP and CYC did not show any potentiation of the antibiotics in the panel, indicating that a covalent connection between all three domains is essential for adjuvant properties.

These studies indicate that TOB-based derivatives containing multiple TOB moieties are potent OMPs that can overcome intrinsic resistance in *P. aeruginosa* against selected antibiotics. However, it is essential to remember that studying has some limitations. Firstly, the cytotoxicity of the new derivatives should be studied on primary kidney cells to measure the release of specific cytotoxic biomarkers. Primary kidney cells are better indicators of nephrotoxicity. Secondly, triple hybrid procedures demand detailed matching of the PK effects of the respective members. As aztreonam (ATM) with avibactam (AVI) combinations have already been optimized for clinical trials, the PK properties of the TOB-TOB-CYP derivative [180] have yet to be investigated.

Bacteria have developed antibiotic resistance mechanisms, including β-lactamases overexpression, porin suppression, outer membrane impermeability, efflux pump overexpression, and target modifications. β-lactamase inhibitors are united with β-lactam antibiotics to preclude antibiotic inactivation by β-lactamases. Gram-negative bacteria are known for their ability to resist treatment with antibiotics, making them a serious threat to human health. One approach to overcoming this resistance is to combine an outer membrane permeabilizer drug with β-lactam antibiotics. This combination works by weakening the protective outer membrane of the bacteria, making it easier for the antibiotics to penetrate and kill the bacterial cells. This approach is particularly effective against pathogens like *P. aeruginosa*, which have outer membranes that are especially difficult to penetrate. This combination therapy increases the effectiveness of antibiotics against these hard-to-treat infections.

Earlier investigations have demonstrated that outer membrane permeabilizers can be developed by fusing tobramycin and nebramine units as dimers. Outer membrane permeabilizers based on trimeric tobramycin and nebramine have been created and shown on a main 1,3,5-triazine framework. The consequent trimers can increase outer membrane antibiotics. Also, it can improve β-lactams and β-lactam/β-lactamase inhibitor hybrids versus resistant *P. aeruginosa* isolates. Three potent trimeric analogs, comprising three TOB and one NEB, were synthesized and rigorously tested for their potentiation effects with various antibiotics, including OM-permeable and OM-impermeable types. The antibiotics, including RIF, NOV, tetracyclines, fluoroquinolones, and BLs, were tested versus Gram-negative bacteria, including *P. aeruginosa*. The outcomes of the analysis indicate that the length of the linker molecule affects the potency of antibiotics in the trimers. The best potentiation occurs when a hydrophobic C12 ether is present, as in the compounds HA and HB (Fig. **50**). The potentiation of BLs and BL/BLI combinations used to treat *P. aeruginosa* infections is particularly interesting. Compounds HA or HB significantly lowered the MIC of BLs and BL/BLI combinations. This lowering was so substantial that microbiological exposure breakpoints of imipenem (IMI), CAZ, and ceftolozame (CTZ) were accomplished by a triple combination of CAZ/AVI, ATM/AVI, and IMI/REL with compounds HA or HB all experimented BL-resistant *P. aeruginosa* isolates [181].

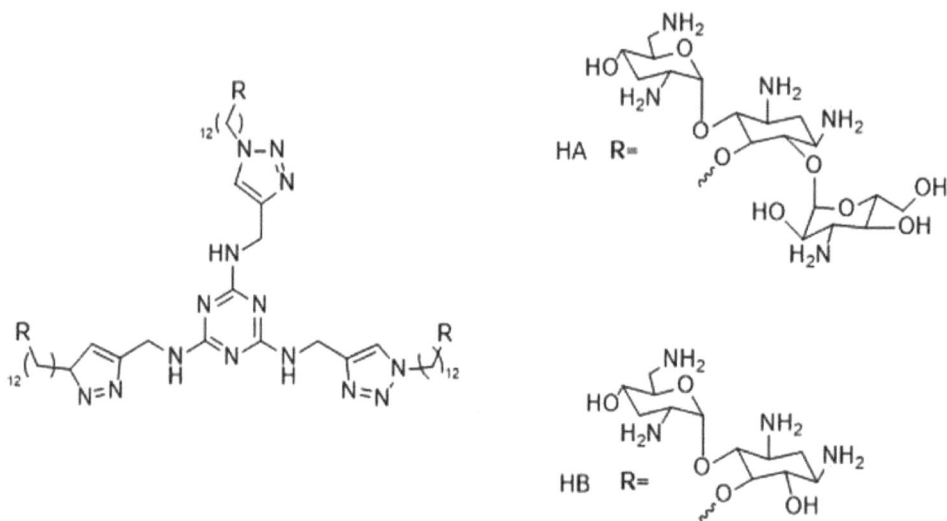

Fig. (50). Chemical structures of the trimeric HA and HB compounds.

NANOPARTICLES

The spread of resistant organisms is the primary concern when developing antimicrobial resistance. New technologies are replacing conventional antimicrobials. Nanotechnology innovations offer hope for patients and practitioners in dealing with pharmaceutical opposition. Nanomaterials have extraordinary possibilities in medical and veterinary specializations [182].

Fighting antibiotic resistance is of utmost importance for the health of both humans and animals. Various strategies are being used to tackle this issue, such as decreasing the excessive utilization of antimicrobials, gathering and evaluating data, preventing the inappropriate use of antimicrobials in livestock, and creating new drugs and nanotechnology. Recent advancements in nanotechnology have resulted in the development of organic and inorganic molecules at the nanoscale level. These tiny particles are anticipated to have practical applications in various industries, such as medicine, therapeutics, food packaging, textiles, and more. One promising area of research is the creation of innovative nanoscale antimicrobial agents and nanocomposites. Such inventions could provide an alternative solution to antimicrobial resistance [183].

Creating successful nanomaterials requires in-depth knowledge of nanoparticles' physicochemical properties and microorganisms' biological aspects. However, addressing the risks associated with using nanoparticles in healthcare is essential. Nanoparticles (NPs) are used to counteract bacteria through various methods, each with a distinct mode of action (Table **12**). Bacterial strains can stick to natural or artificial surfaces and even create biofilms. The factors contributing to bacterial adhesion and biofilm formation include the production of slime-like substances, electrostatic, dipole-dipole, and H-bond interactions, hydrophobicity, and van der Waals forces. Therefore, nanomaterials used as antimicrobial agents should aim to reduce microbial adhesion and biofilm formation. NPs screened for their anti-adhesion capabilities have a higher potential as antimicrobial agents [182].

Table 12. Distinct modes of action of bactericidal nanoparticles.

Type of Nanoparticle	Susceptible Microbes	Mode of Action
Silver (Ag) nanoparticles.	Methicillin-resistant *Staphylococcus aureus*, *Staphylococcus epidermidis*. Vancomycin-resistant *Enterococcus faecium and Klebsiella pneumoniae*.	Inhibit DNA replication and the respiratory chain in bacteria and fungi.Interfere with the electron transport chain and transfer of energy through the membrane.

(Table 12) cont.....

Type of Nanoparticle	Susceptible Microbes	Mode of Action
Carbon-based nanoparticles.	*E. coli, Salmonella enteric, E. faecium, Streptococcus spp., Shewanella oneidensis, Acinetobacter baumannii, Burkholderia cepacia, Yersinia pestis,* and *K. pneumonia.*	Severe damage to the bacterial membrane, physical interaction, inhibition of energy metabolism, and impairment of the respiratory chain.
Magnesium oxide (MgO) nanoparticles.	*S. aureus, E. coli, Bacillus megaterium, Bacillus subtilis*	Formation of reactive oxygen species (ROS), lipid peroxidation, electrostatic interaction, and alkaline effect.
Bismuth (Bi) nanoparticles.	Multiple-antibiotic-resistant *Helicobacter pylori.*	Alter the Krebs cycle, as well as amino acid and nucleotide metabolism.
Titanium dioxide (TiO$_2$) nanoparticles.	*E. coli, S. aureus*, and also against fungi.	Formation of superoxide radicals, ROS, and site-specific DNA damage.
Iron-containing nanoparticles.	*S. aureus, S. epidermidis*, and *E. coli.*	Through ROS-generated oxidative stress. ROS, superoxide radicals (O^{2-}), singlet oxygen (^{1}O2), hydroxyl radicals (OH$^-$), and hydrogen peroxide (H$_2$O$_2$).
Zinc oxide (ZnO) nanoparticles.	*E. coli, Listeria monocytogenes, Salmonella*, and *S. Aureus.*	Hydrogen peroxide generated on the surface of ZnO penetrates the bacterial cells and effectively inhibits growth. Zn2+ ions released from the nanoparticles damage the cell membrane and interact with intracellular components.
Copper oxide (CuO) nanoparticles.	*B. subtilis, S. aureus*, and *E. coli.*	Reduce bacteria at the cell wall. Disrupt the biochemical processes inside bacterial cells.
Gold (Au) nanoparticles.	Methicillin-resistant *S. aureus.*	Generate holes in the cell wall. Bind to the DNA and inhibit the transcription process.
Aluminum (Al) nanoparticles.	*E. coli*	Disrupt cell walls through ROS.

Gold nanoparticles (AuNPs) have been found to possess bactericidal properties, making them ideal for antibacterial applications. Recent studies have shown that bimetallic NPs consisting of gold and other metals, such as Rhodium (Rh) and Ruthenium (Ru), can efficiently inhibit the growth of Gram-negative bacteria, including multi-drug resistant (MDR) strains. The antibacterial properties of these NPs can be regulated by adjusting the ratio of their metal components. Conversely, monometallic NPs such as Au and Rh NPs show no inhibitory effects on *E. coli*. Similarly, mixing Au and Rh NPs does not result in antibacterial activity. However, the combination of AuRh bimetallic NPs can suppress the growth of *E. coli*. Similarly, mixing Au and Ru NPs does not produce any antibiotic effect, yet AuRu NPs can inhibit the proliferation of *E. coli* [184].

The formation of biofilm by bacteria has made it difficult to treat infections due to antibiotic resistance. One such infection caused by bacterial biofilm is a urinary tract infection. Therefore, a recent study aimed to explore alternative methods to inhibit uropathogenic bacteria using nanotechnology. The study isolated Uropathogenic *Escherichia coli* (UPEC) from 110 clinical samples and tested their ability to form biofilm using a Microtiter plate and Congo red agar. The researchers also determined the antibiotic susceptibility of the isolates to identify multidrug-resistant strains. They synthesized silver nanoparticles (AgNPs) using an alcoholic extract of *Lepidium meyenii* yellow root, an eco-friendly and green method. The silver nanoparticles were characterized using several techniques, and they were found to have a diameter of 44.89 nm. The researchers tested the anti-biofilm activity of AgNPs alone and with antibiotics. The scanning electron microscopy observations showed that 7.1825 mg/ml AgNPs prevented biofilm formation. This study is the first to demonstrate the effectiveness of silver nanoparticles synthesized from the alcoholic root extract of *Lepidium meyenii* against UPEC biofilm [185].

Biodegradable Nanoparticles

Biodegradability is the breaking down of nanoparticles (NPs) internally and removing them from the body. It is crucial to clear NPs from the body after they have served their purpose to avoid toxic side effects caused by their accumulation in specific cells, such as those found in the liver and spleen.

The development of biodegradable nanoparticles (BNPs) has revolutionized medicine. These submicron-sized colloidal particles can embed or encapsulate therapeutic agents within their polymeric matrix, adsorb, or conjugate them onto their surface. BNPs' characteristics, such as their bioavailability, suitability for controlled release, high encapsulation capacity, and low toxicity, make them ideal for delivering vaccines, drugs, and other bioactive molecules in a specific area.

Various materials, such as proteins, polysaccharides, and synthetic biodegradable polymers, synthesize biodegradable nanoparticles (BNPs). The selection of polymers depends on the intended use of the nanoparticles, such as biodegradability, biological compatibility, surface functionality, desired size, and the properties of the drug to be encapsulated. Several techniques are available for the synthesis of BNPs.

Materials, including natural compounds such as proteins and polysaccharides and synthetic biodegradable polymers, are used to synthesize biodegradable nanoparticles (BNPs). The nanoparticle's configuration and end application will condition the polymers' selection, such as biodegradability, biological compatibility, surface functionality, desired size, and encapsulated drug properties. Several techniques exist for the synthesis of biodegradable nanoparticles (BNPs). Additionally, it includes the dispersion of preformed polymers, the gelation method, and the´ polymerization of monomers. Various biodegradable polymers, such as poly-ε-caprolactones (PCL), poly-D, L-lactid-co-glycolides (PLGA), and polylactic acids (PLA), are used in the synthesis of BNPs. PLGA, through hydrolysis in the body, produces lactic and glycolic acids, making it ideal for nano vaccines, gene delivery systems, and protein- and peptide-based nanomedicines [186].

PLA is biocompatible and biodegradable and can be prepared through several methods, such as solvent displacement, evaporation, diffusion, and salting-out. PCL could be a polymer for the preparation of BNP, which would have applications in long-term implantable devices and drug delivery. Various techniques, such as solvent displacement/evaporation and nanoprecipitation, can be used to prepare PCL NPs. Gelatin NPs can be prepared by emulsion or desolvation/coacervation methods and are non-toxic, biodegradable, and bioactive.

Poly-alkyl-cyano-acrylates (PAC) polymers produce toxic compounds upon biodegradation, damaging or stimulating the central nervous system and making them unsuitable for humans [187].

Azithromycin (AZI), a macrolide antibiotic, was encapsulated into poly (lactic-c-glycolic acid) (PLGA) polymer using the nano-precipitation method. The effectiveness of AZI-PLGA NPs as an antibacterial agent was tested against Methicillin-resistant *Staphylococcus aureus* (MRSA) and *Enterococcus faecalis* (*E. faecalis*), both resistant to AZI. The study showed that the efflux mechanism was one of the primary reasons for this resistance. The research further revealed that AZI-PLGA NPs were safer than free AZI, as confirmed by the cytotoxicity test. The study yielded that AZI-PLGA NPs can surmount the efflux-resistant mechanism by diminishing the minimum inhibitory concentration (MIC) of AZI-PLGA NPs fourfold compared to free AZI. Specifically, the MIC value for MRSA decreased from 256 to 64 μg/mL, while *E. faecalis* decreased from >1000 to 256 μg/mL. This indicates that AZI-PLGA NPs can be considered a potential therapeutic option for drug-resistant bacterial infections. Further research is needed to explore how the particle size, surface charge, and material composition of PNPs impact their antibacterial activity. It is also essential to ensure the safety of these PNPs, determine their suitability for large-scale manufacture, and explore the possibility of extending this concept to other antibiotics [188].

The encapsulation of antibiotics, such as rifampicin, within biodegradable nanoparticles presents a range of benefits in contrast to free drug administration. These benefits include the reduction of dosages due to localized targeting and sustained release, which subsequently lowers the potential for systemic drug toxicity. The controlled release of the antibacterial agent is made possible through nanoparticles, which provide a barrier between the drug and the surrounding environment. Consequently, this approach allows for a more efficient and targeted delivery of antibiotics while minimizing the potential drug toxicity and adverse effects. Nevertheless, new nanoformulations must be tested in complex biological systems to fully understand their potential to improve drug therapy.

That infectious disease is caused by *Mycobacterium tuberculosis* and requires pervasive and expensive treatment. Incomplete therapy is a significant contributor to the increasing incidence of drug resistance, posing a public health challenge. A recent study suggests that combining antibiotics with bacterial efflux pump inhibitors, such as thioridazine, could improve conventional therapy for tuberculosis. However, the clinical use of thioridazine is problematic due to its toxicity. Therefore, thioridazine was encapsulated in poly(lactic-co-glycolic) acid nanoparticles to address this concern. Experiments were conducted to test how effective thioridazine is when combined with rifampicin nanoparticles or free rifampicin. These combinations were tested on macrophages and a zebrafish model of tuberculosis. Unfortunately, free thioridazine was highly toxic to cells and zebrafish embryos. However, when thioridazine was encapsulated in nanoparticles, no toxicity was detected. Combining rifampicin nanoparticles with thioridazine nanoparticles resulted in a moderate increase in the killing of *Mycobacterium bovis* BCG *and M. tuberculosis* in macrophages. The combination of thioridazine nanoparticles with rifampicin showed significant therapeutic effects in zebrafish. This combination enhanced embryo survival and reduced mycobacterial infection. These studies show that zebrafish embryos are vulnerable to drug toxicity. It also confirms that thioridazine nanoparticle therapy can improve the antibacterial impact of rifampicin *in vivo* [189].

Antimicrobial peptides (AMPs) show potential as alternatives to antibiotics and as adjuvants that can improve antibiotic efficacy. Additionally, biodegradable lipid nanoparticles could increase the antibacterial activity of antibiotics and antimicrobial peptides. This investigation studied the interaction of nisin Z and melittin, two AMPs, with conventional antibiotics against *Staphylococcus aureus, Staphylococcus epidermidis*, and *E. coli*. It also evaluates the effectiveness of nanostructured lipid carriers (NLCs) in entrapping nisin Z. The study's findings demonstrate that nisin Z exhibits additive interactions with various conventional antibiotics, with notable synergism observed for the novobiocin-nisin Z combinations.

Furthermore, it was observed that the effectiveness of free nisin Z as an antimicrobial agent against *E. coli* could be enhanced by utilizing the non-antibiotic adjuvant EDTA. Additionally, the presence of nisin Z in nanostructured lipid carriers (NLCs) was effective against Gram-positive species at physiological pH, and their effectiveness was further improved in the presence of EDTA. These findings suggest that nisin Z could be a promising adjuvant in antimicrobial therapy and support the fight against antibiotic resistance. The study focuses on using biodegradable lipid nanoparticles, which can heighten the antibacterial activity of antibiotics and antimicrobial peptides. NLCs, conversely, can enhance nisin Z's antibacterial activity towards Gram-positive bacterial species commonly associated with skin infections [190].

Intracellular bacteria are responsible for causing various severe illnesses worldwide. Some examples of these bacteria are *Mycobacterium* spp., *Brucella* spp., and *Listeria monocytogenes*. These bacteria have developed several clever methods to exploit host processes, which allow them to multiply and spread without harming the host cells. This enables them to maintain their intracellular lifestyle. They are resulting in persistent, severe, and latent infections. Access to antimicrobial agents in infected cells and intracellular niches where pathogens reside with sufficiently high therapeutic concentrations is necessary to treat intracellular infections effectively. However, treating intracellular infections is a great challenge due to poor cell penetration, limited intracellular retention, unsatisfactory subcellular distribution, and decreased intracellular activity of most antimicrobial agents.

Due to their high polarity, aminoglycosides such as gentamicin penetrate very slowly through cell membranes and, once inside, remain confined to endolysosomal compartments, where the endosomal acidic pH reduces their activity. A hydrophobically modified gentamicin, gentamicin-AOT [AOT is sodium salt of bis(2-ethylhexyl) sulfosuccinate], was encapsulated in poly(lactic-co-glycolic acid) (PLGA) nanoparticles. Over 24 hours, the accumulation of free gentamicin-AOT continued linearly and reached an apparent intracellular-t--extracellular concentration ratio of 3 without any plateau. After being incubated for 24 hours, the 752HPLGA nanoparticles proved effective in accumulating ten times more gentamicin in phagocytic cells than free gentamicin. The gentamicin accumulated in the cytosol with a predominant subcellular localization. This result encourages additional studies to achieve the clinical use of gentamicin-AOT/PLGA nanoparticles [191].

Intracellular delivery requires using materials that can bind to and recognize the eukaryotic cell membrane. Passive strategies take advantage of recognition by the immune system, as macrophages can naturally remember and internalize NPs through an energy-dependent phagocytic mechanism. Therefore, the antimicrobial payload of NPs can accumulate within the intracellular compartments of infected cells, fusing with pathogen-containing vesicles or releasing the NP payload into the cytoplasm.

An administration system has been designed so that there is a possible colocalization between the nanoparticle loaded with the drug and the bacteria in the same intracellular compartment, improving antimicrobial efficacy. Nanoparticulate formulations of PLGA (Resomer RG504) NPs loaded with cloxacillin were used against methicillin-susceptible (MSSA) and resistant (MRSA) *S. aureus* compared to the corresponding concentration of the free drug.

The results indicated that drug-loaded NPs significantly reduced intracellular bacterial load in cellular infection models. Therefore, this novel delivery system shows excellent potential for successful targeted treatment of staphylococcal infections due to its intracellular drug delivery, release capabilities, and high cytocompatibility [192].

CONCLUSION

The phenomenon of bacterial resistance to antibiotics has been known for a long time. However, it had not become a public health problem. Unquestionably, bacterial resistance to antibiotics occurred as a natural response of bacteria to the aggression that humanity exerted against them. Point mutations allowed some bacteria to show resistance to some bactericides. This genetic information was transmitted vertically to their offspring. In a similar exercise, these bacteria suffered mutations that made them resistant to a second antibiotic, resulting in bacteria that were resistant to two bactericides. This genetic information can be transmitted vertically and horizontally through plasmids or transposons, among other mechanisms. Horizontal transmission allows antibiotic resistance gene information to be shared within the same species and between bacteria of different species. Unquestionably, these vertical and horizontal mechanisms give bacteria crucial genetic information to resist bactericides.

This resistance gene information is transmitted to bacteria in different ways, such as cell wall change, porin closure, or the synthesis of enzymes that destroy or disable antibiotics, such as β-lactamases. This means that, nowadays, strains that are multi-resistant to bactericides can occur.

Given this panorama, WHO and other organizations have been alerting the general population to the problem and encouraging the scientific community to synthesize new bactericides with new mechanisms of activity.

Although the problem of bacterial resistance is severe, some aspects can be used to confront this threat. One of them is bacterial resistance in certain strains of a particular species. For example, certain strains of *Pseudomonas aeruginosa* have β-lactamases, but others of the same species lack them. This makes it possible to study both strains and find some ways to combat them. For example, recently, drug combinations have increased to attack bacteria by at least two routes or the synthesis of so-called hybrids, which is the covalent union of at least two drugs. It is clear that using combinations Dis or hybrids is not the final solution, but it will give time to find a better solution.

Another exciting aspect is the so-called antivirulence therapy. This therapy is based on the quorum sensing system (QS) linked to the bacterial population density. As its name indicates, the main objective of this therapy is to inhibit the so-called virulence factors, but without eliminating the bacteria, which will not create selection pressure, and bacterial resistance will be avoided, at least theoretically. So far, there is no QS-based therapy in the clinic, which indicates that it must be studied in real scenarios to see if the results obtained in laboratory conditions can be applied in actual infection situations.

ACKNOWLEDGEMENTS

The author thanks Ms. Graciela Flores-Rosete, whose assistance was fundamental to the completion of this study.

REFERENCES

[1]   Blum HE. The human microbiome. Adv Med Sci 2017; 62(2): 414-20.
        [http://dx.doi.org/10.1016/j.advms.2017.04.005] [PMID: 28711782]

[2]   Wagner DM, Klunk J, Harbeck M, *et al. Yersinia pestis* and the Plague of Justinian 541–543 AD: a genomic analysis. Lancet Infect Dis 2014; 14(4): 319-26.
        [http://dx.doi.org/10.1016/S1473-3099(13)70323-2] [PMID: 24480148]

[3]   Sender R, Fuchs S, Milo R. Revised estimates for the number of human and bacteria cells in the body. PLoS Biol 2016; 14(8): e1002533.
        [http://dx.doi.org/10.1371/journal.pbio.1002533] [PMID: 27541692]

[4]   Quigley EMM. The spectrum of small intestinal bacterial overgrowth (SIBO). Curr Gastroenterol Rep 2019; 21(1): 3.
        [http://dx.doi.org/10.1007/s11894-019-0671-z] [PMID: 30645678]

[5]   Structure, function and diversity of the healthy human microbiome. Nature 2012; 486(7402): 207-14.
        [http://dx.doi.org/10.1038/nature11234] [PMID: 22699609]

[6]   Ramsamy Y, Mlisana KP, Amoako DG, *et al.* Mobile genetic elements-mediated Enterobacterales-associated carbapenemase antibiotic resistance genes propagation between the environment and

humans: A One Health South African study. Sci Total Environ 2022; 806(Pt 3): 150641.
[http://dx.doi.org/10.1016/j.scitotenv.2021.150641] [PMID: 34606866]

[7]   Organización Panamericana de la Salud 2019. Tratamiento de las enfermedades infecciosas 2020-2022; Octava edición ISBN: 978-92-75-32100-3 eISBN: 978-92-75-32113-3.

[8]   dos Santos GS, Solidônio EG, Costa MCVV, *et al.* Study of the Enterobacteriaceae Group CESP (Citrobacter, Enterobacter, Serratia, Providencia, Morganella and Hafnia): A Review. The Battle Against Microbial Pathogens: Basic Science, Technological Advances and Educational Programs (A. Méndez-Vilas, Ed.) Publisher: Formatext 2015).

[9]   Wang YC, Tang HL, Liao YC, *et al.* Cocarriage of Distinct bla_{KPC-2} and bla_{OXA-48} Plasmids in a Single Sequence Type 11 Carbapenem-Resistant *Klebsiella pneumoniae* Isolate. Antimicrob Agents Chemother 2019; 63(6): e02282-18.
[http://dx.doi.org/10.1128/AAC.02282-18] [PMID: 30962338]

[10]   Alekshun MN, Levy SB. Molecular mechanisms of antibacterial multidrug resistance. Cell 2007; 128(6): 1037-50.
[http://dx.doi.org/10.1016/j.cell.2007.03.004] [PMID: 17382878]

[11]   National Institutes of Health. National Library of Medicine National Center for Biotechnology Information https://pubchem.ncbi.nlm.nih.gov/compound/Imipenem

[12]   National Institutes of Health, National Library of Medicine. National Center for Biotechnology Information https://pubchem.ncbi.nlm.nih.gov/compound/Meropenem

[13]   Dever LA, Dermody TS. Mechanisms of bacterial resistance to antibiotics. Arch Intern Med 1991; 151(5): 886-95.
[http://dx.doi.org/10.1001/archinte.1991.00400050040010] [PMID: 2025137]

[14]   National Institutes of Health, National Library of Medicine. National Center for Biotechnology Information https://pubchem.ncbi.nlm.nih.gov/compound/Ertapenem

[15]   National Institutes of Health, National Library of Medicine. National Center for Biotechnology Information https://pubchem.ncbi.nlm.nih.gov/compound/Doripenem

[16]   Grundmann H, Livermore DM, Giske CG, *et al.* Carbapenem-non-susceptible Enterobacteriaceae in Europe: conclusions from a meeting of national experts. Euro Surveill 2010; 15(46): 19711.
[http://dx.doi.org/10.2807/ese.15.46.19711-en] [PMID: 21144429]

[17]   Ke W, Bethel CR, Thomson JM, Bonomo RA, van den Akker F. Crystal structure of KPC-2: insights into carbapenemase activity in class A β-lactamases. Biochemistry 2007; 46(19): 5732-40.
[http://dx.doi.org/10.1021/bi700300u] [PMID: 17441734]

[18]   Patel G, Bonomo RA. Status report on carbapenemases: challenges and prospects. Expert Rev Anti Infect Ther 2011; 9(5): 555-70.
[http://dx.doi.org/10.1586/eri.11.28] [PMID: 21609267]

[19]   Kim CH, Jeong YJ, Lee J, *et al.* Essential role of toll-like receptor 4 in Acinetobacter baumannii induced immune responses in immune cells. Microb Pathog 2013; 54: 20-5.
[http://dx.doi.org/10.1016/j.micpath.2012.08.008] [PMID: 22982140]

[20]   Vijayakumar S, Anandan S, Ms DP, *et al.* Insertion sequences and sequence types profile of clinical isolates of carbapenem-resistant *A. baumannii* collected across India over four year period. J Infect Public Health 2020; 13(7): 1022-8.
[http://dx.doi.org/10.1016/j.jiph.2019.11.018] [PMID: 31874816]

[21]   Cruz-Muñiz MY, López-Jacome LE, Hernández-Durán M, *et al.* Repurposing the anticancer drug mitomycin C for the treatment of persistent *Acinetobacter baumannii* infections. Int J Antimicrob Agents 2017; 49(1): 88-92.
[http://dx.doi.org/10.1016/j.ijantimicag.2016.08.022] [PMID: 27939675]

[22]   Kostyanev T, Xavier BB, García-Castillo M, *et al.* Phenotypic and molecular characterizations of

carbapenem-resistant *Acinetobacter baumannii* isolates collected within the EURECA study. Int J Antimicrob Agents 2021; 57(6): 106345.
[http://dx.doi.org/10.1016/j.ijantimicag.2021.106345] [PMID: 33887390]

[23]   Medioli F, Bacca E, Faltoni M, *et al.* Is It Possible to Eradicate Carbapenem-Resistant *Acinetobacter baumannii* (CRAB) from Endemic Hospitals? Antibiotics (Basel) 2022; 11(8): 1015.
[http://dx.doi.org/10.3390/antibiotics11081015] [PMID: 36009885]

[24]   Wong D, Nielsen TB, Bonomo RA, Pantapalangkoor P, Luna B, Spellberg B. Clinical and Pathophysiological Overview of Acinetobacter Infections: a Century of Challenges. Clin Microbiol Rev 2017; 30(1): 409-47.
[http://dx.doi.org/10.1128/CMR.00058-16] [PMID: 27974412]

[25]   Tripodi MF, Durante-Mangoni E, Fortunato R, Utili R, Zarrilli R. Comparative activities of colistin, rifampicin, imipenem and sulbactam/ampicillin alone or in combination against epidemic multidrug-resistant *Acinetobacter baumannii* isolates producing OXA-58 carbapenemases. Int J Antimicrob Agents 2007; 30(6): 537-40.
[http://dx.doi.org/10.1016/j.ijantimicag.2007.07.007] [PMID: 17851050]

[26]   Cai Y, Chai D, Wang R, Liang B, Bai N. Colistin resistance of *Acinetobacter baumannii*: clinical reports, mechanisms and antimicrobial strategies. J Antimicrob Chemother 2012; 67(7): 1607-15.
[http://dx.doi.org/10.1093/jac/dks084] [PMID: 22441575]

[27]   Stein GE, Babinchak T. Tigecycline: an update. Diagn Microbiol Infect Dis 2013; 75(4): 331-6.
[http://dx.doi.org/10.1016/j.diagmicrobio.2012.12.004] [PMID: 23357291]

[28]   Ruggiero A, Cappelletti F, Bettua C, *et al.* Tigecycline in the treatment of multidrug-resistant *Acinetobacter baumannii*: A real-life multicenter experience from the Italian Society of Anti-Infective Therapy (SITA) network. Int J Antimicrob Agents 2021; 58(1): 106273.
[http://dx.doi.org/10.1016/j.ijantimicag.2021.106273]

[29]   Ni W, Han Y, Zhao J, *et al.* Tigecycline treatment experience against multidrug-resistant *Acinetobacter baumannii* infections: a systematic review and meta-analysis. Int J Antimicrob Agents 2016; 47(2): 107-16.
[http://dx.doi.org/10.1016/j.ijantimicag.2015.11.011] [PMID: 26742726]

[30]   Duncan LR, Wang W, Sader HS. *In vitro* potency and spectrum of the novel polymyxin MRX-8 tested against clinical isolates of gram-negative bacteria. Antimicrob Agents Chemother 2022; 66(5): e00139-22.
[http://dx.doi.org/10.1128/aac.00139-22] [PMID: 35475635]

[31]   Velkov T, Roberts KD, Thompson PE, Li J. Polymyxins: a new hope in combating Gram-negative superbugs? Future Med Chem 2016; 8(10): 1017-25.
[http://dx.doi.org/10.4155/fmc-2016-0091] [PMID: 27328129]

[32]   Jiang X, Yang K, Han ML, *et al.* Outer Membranes of Polymyxin-Resistant *Acinetobacter baumannii* with Phosphoethanolamine-Modified Lipid A and Lipopolysaccharide Loss Display Different Atomic-Scale Interactions with Polymyxins. ACS Infect Dis 2020; 6(10): 2698-708.
[http://dx.doi.org/10.1021/acsinfecdis.0c00330] [PMID: 32871077]

[33]   Isler B, Doi Y, Bonomo RA, Paterson DL. New Treatment Options against Carbapenem-Resistant *Acinetobacter baumannii* Infections. Antimicrob Agents Chemother 2019; 63(1): e01110-18.
[http://dx.doi.org/10.1128/AAC.01110-18] [PMID: 30323035]

[34]   Monogue ML, Tsuji M, Yamano Y, Echols R, Nicolau DP. Efficacy of Humanized Exposures of Cefiderocol (S-649266) against a Diverse Population of Gram-Negative Bacteria in a Murine Thigh Infection Model. Antimicrob Agents Chemother 2017; 61(11): e01022-17.
[http://dx.doi.org/10.1128/AAC.01022-17] [PMID: 28848004]

[35]   Solomkin J, Evans D, Slepavicius A, *et al.* Assessing the efficacy and safety of eravacycline *vs* ertapenem in complicated intra-abdominal infections in the investigating gram-Negative infections treated with eravacycline (IGNITE 1) trial. JAMA Surg 2017; 152(3): 224-32.

[http://dx.doi.org/10.1001/jamasurg.2016.4237] [PMID: 27851857]

[36]   Bassetti M, Righi E. Eravacycline for the treatment of intra-abdominal infections. Expert Opin Investig Drugs 2014; 23(11): 1575-84.
[http://dx.doi.org/10.1517/13543784.2014.965253] [PMID: 25251475]

[37]   Li Y, Cui L, Xue F, Wang Q, Zheng B. Synergism of eravacycline combined with other antimicrobial agents against carbapenem-resistant Enterobacteriaceae and *Acinetobacter baumannii*. J Glob Antimicrob Resist 2022; 30: 56-9.
[http://dx.doi.org/10.1016/j.jgar.2022.05.020] [PMID: 35660472]

[38]   Van Hise N, Petrak RM, Skorodin NC, *et al.* A Real-World Assessment of Clinical Outcomes and Safety of Eravacycline: A Novel Fluorocycline. Infect Dis Ther 2020; 9(4): 1017-28.
[http://dx.doi.org/10.1007/s40121-020-00351-0] [PMID: 33063176]

[39]   Zhanel GG, Cheung D, Adam H, *et al.* Review of Eravacycline, a Novel Fluorocycline Antibacterial Agent. Drugs 2016;76:567–588. https://doi:10.1007/s40265-016-0545-8.

[40]   Shaeer KM, Zmarlicka MT, Chahine EB, Piccicacco N, Cho JC. Plazomicin: A Next-Generation Aminoglycoside. Pharmacotherapy 2019; 39(1): 77-93.
[http://dx.doi.org/10.1002/phar.2203] [PMID: 30511766]

[41]   Castanheira M, Sader HS, Mendes RE, Jones RN. Activity of Plazomicin Tested against *Enterobacterales* Isolates Collected from U.S. Hospitals in 2016–2017: Effect of Different Breakpoint Criteria on Susceptibility Rates among Aminoglycosides. Antimicrob Agents Chemother 2020; 64(5): e02418-19.
[http://dx.doi.org/10.1128/AAC.02418-19] [PMID: 32094137]

[42]   Cloutier DJ, Komirenko AS, Cebrik DS, *et al.* Plazomicin vs. Meropenem for complicated urinarytract infection (cUTI) and acute pyelonephritis (AP): diagnosis-specific results from the phase 3 EPIC study. Open Forum Infect Dis 2017; 4(I.suppl_1): S532.
[http://dx.doi.org/10.1093/ofid/ofx163.1385] [PMID: 30668657]

[43]   Mirzaei B, Babaei R, Bazgir ZN, Goli HR, Keshavarzi S, Amiri E. Prevalence of Enterobacteriaceae spp. and its multidrug-resistant rates in clinical isolates: A two-center cross-sectional study. Mol Biol Rep 2021; 48(1): 665-75.
[http://dx.doi.org/10.1007/s11033-020-06114-x] [PMID: 33389531]

[44]   Bassetti M, Giacobbe DR, Giamarellou H, *et al.* Critically Ill Patients Study Group of the European Society of Clinical Microbiology and Infectious Disease (ESCMID); Hellenic Society of Chemotherapy (HSC) and Società Italiana di Terapia Antinfettiva (SITA). Management of KPC-producing *Klebsiella pneumoniae* infections. Clin Microbiol Infect 2018; 24(2): 133-44.
[http://dx.doi.org/10.1016/j.cmi.2017.08.030] [PMID: 28893689]

[45]   Bassetti M, Poulakou G, Ruppe E. Characteristics of carbapenemase-producing Enterobacteriaceae in Europe: Epidemiology, clinical and laboratory features, detection and treatment. Clin Microbiol Infect 2017; 23(10): 809-16.
[http://dx.doi.org/10.1016/j.cmi.2017.03.017]

[46]   http://www.cdc.gov/hai/organisms/cre/definition.html

[47]   Reyes JA, Melano R, Cárdenas PA, Trueba G. Mobile genetic elements associated with carbapenemase genes in South American Enterobacterales. Braz J Infect Dis 2020; 24(3): 231-8.
[http://dx.doi.org/10.1016/j.bjid.2020.03.002] [PMID: 32325019]

[48]   Tischendorf J, de Avila RA, Safdar N. Risk of infection following colonization with carbapenem-resistant Enterobactericeae: A systematic review. Am J Infect Control 2016; 44(5): 539-43.
[http://dx.doi.org/10.1016/j.ajic.2015.12.005] [PMID: 26899297]

[49]   Parker JK, Gu R, Estrera GA, *et al.* Carbapenem-Resistant and ESBL-Producing Enterobacterales Emerging in Central Texas. Infect Drug Resist 2023; 16: 1249-61.
[http://dx.doi.org/10.2147/IDR.S403448] [PMID: 36891378]

[50]   Bratu S, Tolaney P, Karumudi U, *et al.* Carbapenemase-producing *Klebsiella pneumoniae* in Brooklyn, NY: molecular epidemiology and *in vitro* activity of polymyxin B and other agents. J Antimicrob Chemother 2005; 56(1): 128-32.
[http://dx.doi.org/10.1093/jac/dki175] [PMID: 15917285]

[51]   Olsson A, Hong M, Al-Farsi H, Giske CG, Lagerbäck P, Tängdén T. Interactions of polymyxin B in combination with aztreonam, minocycline, meropenem, and rifampin against *Escherichia coli* producing NDM and OXA-48-group carbapenemases. Antimicrob Agents Chemother 2021; 65(12): e01065-21.
[http://dx.doi.org/10.1128/AAC.01065-21] [PMID: 34516251]

[52]   Samonis G, Maraki S, Karageorgopoulos DE, Vouloumanou EK, Falagas ME. Synergy of fosfomycin with carbapenems, colistin, netilmicin, and tigecycline against multidrug-resistant *Klebsiella pneumoniae*, *Escherichia coli*, and *Pseudomonas aeruginosa* clinical isolates. Eur J Clin Microbiol Infect Dis 2012; 31(5): 695-701.
[http://dx.doi.org/10.1007/s10096-011-1360-5] [PMID: 21805292]

[53]   Panagiotakopoulou A, Daikos GL, Miriagou V, Loli A, Tzelepi E, Tzouvelekis LS. Comparative in vitro killing of carbapenems and aztreonam against *Klebsiella pneumoniae* producing VIM-1 metallo-β-lactamase. Int J Antimicrob Agents 2007; 29(3): 360-2.
[http://dx.doi.org/10.1016/j.ijantimicag.2006.11.004] [PMID: 17223016]

[54]   Aydemir Ö, Şahin EÖ, Ayhancı T, *et al.* Investigation of *In-vitro* Efficacy of Intravenous Fosfomycin in Extensively Drug-Resistant *Klebsiella pneumoniae* Isolates and Effect of Glucose 6-Phosphate on Sensitivity Results. Int J Antimicrob Agents 2022; 59(1): 106489.
[http://dx.doi.org/10.1016/j.ijantimicag.2021.106489] [PMID: 34848325]

[55]   Tutone M, Bjerklund Johansen TE, Cai T, Mushtaq S, Livermore DM. Susceptibility and Resistance to Fosfomycin and other antimicrobial agents among pathogens causing lower urinary tract infections: findings of the SURF study. Int J Antimicrob Agents 2022; 59(5): 106574.
[http://dx.doi.org/10.1016/j.ijantimicag.2022.106574] [PMID: 35307561]

[56]   Rabaan AA, Eljaaly K, Alfouzan WA, *et al.* Psychogenetic, genetic and epigenetic mechanisms in *Candida auris*: Role in drug resistance. J Infect Public Health 2023; 16(2): 257-63.
[http://dx.doi.org/10.1016/j.jiph.2022.12.012] [PMID: 36608452]

[57]   Pristov KE, Ghannoum MA. Resistance of Candida to azoles and echinocandins worldwide. Clin Microbiol Infect 2019; 25(7): 792-8.
[http://dx.doi.org/10.1016/j.cmi.2019.03.028] [PMID: 30965100]

[58]   Rybak JM, Barker KS, Munoz JF, *et al. In vivo* emergence of high-level resistance during treatment reveals the first identified mechanism of amphotericin B resistance in *Candida auris*. Clinical Microbiology and Infection 2022; 28(6): 838e843.

[59]   Raschig M, Ramírez-Zavala B, Wiest J, *et al.* Azobenzene derivatives with activity against drug-resistant *Candida albicans* and *Candida auris*. Arch Pharm (Weinheim) 2023; 356(2): 2200463.
[http://dx.doi.org/10.1002/ardp.202200463] [PMID: 36403201]

[60]   Singh S, Barbarino A, Youssef EG, Coleman D, Gebremariam T, Ibrahim AS. Protective Efficacy of Anti-Hyr1p Monoclonal Antibody against Systemic Candidiasis Due to Multi-Drug-Resistant *Candida auris*. J Fungi (Basel) 2023; 9(1): 103.
[http://dx.doi.org/10.3390/jof9010103] [PMID: 36675924]

[61]   Parker RA, Gabriel KT, Graham KD, Butts BK, Cornelison CT. Antifungal Activity of Select Essential Oils against *Candida auris* and Their Interactions with Antifungal Drugs. Pathogens 2022; 11(8): 821.
[http://dx.doi.org/10.3390/pathogens11080821] [PMID: 35894044]

[62]   Ellsworth M, Ostrosky-Zeichner L. Isavuconazole: Mechanism of Action, Clinical Efficacy, and Resistance. J Fungi (Basel) 2020; 6(4): 324.
[http://dx.doi.org/10.3390/jof6040324] [PMID: 33260353]

[63]   Maertens JA, Raad II, Marr KA, *et al.* Isavuconazole *versus* voriconazole for primary treatment of invasive mould disease caused by Aspergillus and other filamentous fungi (SECURE): a phase 3, randomised-controlled, non-inferiority trial. Lancet 2016; 387(10020): 760-9.
[http://dx.doi.org/10.1016/S0140-6736(15)01159-9] [PMID: 26684607]

[64]   Zhang T, Shen Y, Feng S. Clinical research advances of isavuconazole in the treatment of invasive fungal diseases. Front Cell Infect Microbiol 2022; 12: 1049959.
[http://dx.doi.org/10.3389/fcimb.2022.1049959] [PMID: 36530445]

[65]   Shaw KJ, Ibrahim AS. Fosmanogepix: A Review of the First-in-Class Broad Spectrum Agent for the Treatment of Invasive Fungal Infections. J Fungi (Basel) 2020; 6(4): 239.
[http://dx.doi.org/10.3390/jof6040239] [PMID: 33105672]

[66]   Jiménez-Ortigosa C, Perez WB, Angulo D, Borroto-Esoda K, Perlin DS. *De novo* acquisition of resistance to SCY-078 in *Candida glabrata* involves FKS mutations that both overlap and are distinct from those conferring echinocandin resistance. Antimicrob Agents Chemother 2017; 61(9): e00833-17.
[http://dx.doi.org/10.1128/AAC.00833-17] [PMID: 28630180]

[67]   Pfaller MA, Messer SA, Rhomberg PR, Borroto-Esoda K, Castanheira M. Differential activity of the oral glucan synthase inhibitor SCY078 against wild-type and echinocandin-resistant strains of Candida species. Antimicrob Agents Chemother 2017; 61(8): e00161-17.
[http://dx.doi.org/10.1128/AAC.00161-17] [PMID: 28533234]

[68]   Gamal A, Chu S, McCormick TS, Borroto-Esoda K, Angulo D, Ghannoum MA. Ibrexafungerp, a novel oral triterpenoid antifungal in development: Overview of antifungal activity against *Candida glabrata*. Front Cell Infect Microbiol 2021; 11: 642358.
[http://dx.doi.org/10.3389/fcimb.2021.642358] [PMID: 33791244]

[69]   Petraitis V, Petraitiene R, Katragkou A, *et al.* Combination Therapy with Ibrexafungerp (Formerly SCY-078), a First-in-Class Triterpenoid Inhibitor of (1→3)-β- D -Glucan Synthesis, and Isavuconazole for Treatment of Experimental Invasive Pulmonary Aspergillosis. Antimicrob Agents Chemother 2020; 64(6): e02429-19.
[http://dx.doi.org/10.1128/AAC.02429-19] [PMID: 32179521]

[70]   Davis MR, Donnelley MA, Thompson GR III. Ibrexafungerp: A novel oral glucan synthase inhibitor. Med Mycol 2020; 58(5): 579-92.
[http://dx.doi.org/10.1093/mmy/myz083] [PMID: 31342066]

[71]   Arendrup MC, Jørgensen KM, Hare RK, Chowdhary A. *In Vitro* Activity of Ibrexafungerp (SCY-078) against *Candida auris* Isolates as Determined by EUCAST Methodology and Comparison with Activity against *C. albicans* and *C. glabrata* and with the Activities of Six Comparator Agents. Antimicrob Agents Chemother 2020; 64(3): e02136-19.
[http://dx.doi.org/10.1128/AAC.02136-19] [PMID: 31844005]

[72]   Larkin E, Hager C, Chandra J, *et al.* The Emerging Pathogen *Candida auris*: Growth Phenotype, Virulence Factors, Activity of Antifungals, and Effect of SCY-078, a Novel Glucan Synthesis Inhibitor, on Growth Morphology and Biofilm Formation. Antimicrob Agents Chemother 2017; 61(5): e02396-16.
[http://dx.doi.org/10.1128/AAC.02396-16] [PMID: 28223375]

[73]   Caballero U, Kim S, Eraso E, *et al.* *In Vitro* Synergistic Interactions of Isavuconazole and Echinocandins against *Candida auris*. Antibiotics (Basel) 2021; 10(4): 355.
[http://dx.doi.org/10.3390/antibiotics10040355] [PMID: 33800601]

[74]   Aghaei Gharehbolagh S, Izadi A, Talebi M, *et al.* New weapons to fight a new enemy: A systematic review of drug combinations against the drug-resistant fungus *Candida auris*. Mycoses 2021; 64(11): 1308-16.
[http://dx.doi.org/10.1111/myc.13277] [PMID: 33774879]

[75]   O'Grady K, Knight DR, Riley TV. Antimicrobial resistance in *Clostridioides difficile*. Eur J Clin

Microbiol Infect Dis 2021; 40(12): 2459-78.
[http://dx.doi.org/10.1007/s10096-021-04311-5] [PMID: 34427801]

[76] Wickramage I, Spigaglia P, Sun X. Mechanisms of antibiotic resistance of *Clostridioides difficile*. J Antimicrob Chemother 2021; 76(12): 3077-90.
[http://dx.doi.org/10.1093/jac/dkab231] [PMID: 34297842]

[77] Jarmo O, Veli-Jukka A, Eero M. Treatment of *Clostridioides (Clostridium) difficile* infection. Ann Med 2020; 52(1-2): 12-20.
[http://dx.doi.org/10.1080/07853890.2019.1701703] [PMID: 31801387]

[78] Kotila SM, Mentula S, Ollgren J, Virolainen-Julkunen A, Lyytikäinen O. Community and healthcare-associated *Clostridium difficile* infections, Finland, 2008-2013. Emerg Infect Dis 2016; 22(10): 1747-53.
[http://dx.doi.org/10.3201/eid2210.151492] [PMID: 27648884]

[79] Hui W, Li T, Liu W, Zhou C, Gao F. Fecal microbiota transplantation for treatment of recurrent *C. difficile infection*: An updated randomized controlled trial meta-analysis. PLoS One 2019; 14(1): e0210016.
[http://dx.doi.org/10.1371/journal.pone.0210016] [PMID: 30673716]

[80] Johnson S, Gerding DN. Bezlotoxumab. Clin Infect Dis 2019; 68(4): 699-704.
[http://dx.doi.org/10.1093/cid/ciy577] [PMID: 30020417]

[81] Prehn Jv, Reigadas E, Vogelzang EH, *et al.* European Society of Clinical Microbiology and Infectious Diseases: 2021 update on the treatment guidance document for Clostridioides difficile infection in adults. Clinical Microbiology and Infection 2021;Suppl2:(S1-S21). S1eS21.
https://doi.org/10.1016/j.cmi.2021.09.038

[82] Zhanel GG, Walkty AJ, Karlowsky JA. Fidaxomicin: A novel agent for the treatment of *Clostridium difficile* infection. Can J Infect Dis Med Microbiol 2015; 26(6): 305-12.
[http://dx.doi.org/10.1155/2015/934594] [PMID: 26744587]

[83] https://www.who.int/news/item/22-11-2021-gonorrhoea-antimicrobial-resistance-resu-ts-and-guidance-vaccine-developmen

[84] Bradford PA, Miller AA, O'Donnell J, Mueller JP. Zoliflodacin: An Oral Spiropyrimidinetrione Antibiotic for the Treatment of *Neisseria gonorrheae*, Including Multi-Drug-Resistant Isolates. ACS Infect Dis 2020; 6(6): 1332-45.
[http://dx.doi.org/10.1021/acsinfecdis.0c00021] [PMID: 32329999]

[85] Jacobsson S, Paukner S, Golparian D, Jensen JS, Unemo M. *In Vitro* Activity of the Novel Pleuromutilin Lefamulin (BC-3781) and Effect of Efflux Pump Inactivation on Multidrug-Resistant and Extensively Drug-Resistant *Neisseria gonorrhoeae*. Antimicrob Agents Chemother 2017; 61(11): e01497-17.https://doi-org.pbidi.unam.mx:2443/10.1128/AAC.01497-17
[http://dx.doi.org/10.1128/AAC.01497-17] [PMID: 28893785]

[86] Anayo OF, Scholastica EC, Peter OC, *et al.* The Beneficial Roles of Pseudomonas in Medicine, Industries, and Environment: A Review In: Sriramulu D (Ed) Pseudomonas aeruginosa - An Armory Within IntechOpen;. 2019.
[http://dx.doi.org/10.5772/intechopen.85996]

[87] Hirsch EB, Tam VH. Impact of multidrug-resistant *Pseudomonas aeruginosa* infection on patient outcomes. Expert Rev Pharmacoecon Outcomes Res 2010; 10(4): 441-51.
[http://dx.doi.org/10.1586/erp.10.49] [PMID: 20715920]

[88] Kamiya H, Ehara T, Matsumoto T. Inhibitory effects of lactoferrin on biofilm formation in clinical isolates of *Pseudomonas aeruginosa*. J Infect Chemother 2012; 18(1): 47-52.

[89] Silver LL. Multi-targeting by monotherapeutic antibacterials. Nat Rev Drug Discov 2007; 6(1): 41-55.
[http://dx.doi.org/10.1038/nrd2202] [PMID: 17159922]

[90] Covington BC, McLean JA, Bachmann BO. Comparative mass spectrometry-based metabolomics

strategies for the investigation of microbial secondary metabolites. Nat Prod Rep 2017; 34(1): 6-24.
[http://dx.doi.org/10.1039/C6NP00048G] [PMID: 27604382]

[91] Amaral L, Viveiros M. Why thioridazine in combination with antibiotics cures extensively drug-resistant *Mycobacterium tuberculosis* infections. Int J Antimicrob Agents 2012; 39(5): 376-80.
[http://dx.doi.org/10.1016/j.ijantimicag.2012.01.012] [PMID: 22445204]

[92] Lee HJ, Bergen PJ, Bulitta JB, *et al.* Synergistic activity of colistin and rifampin combination against multidrug-resistant *Acinetobacter baumannii* in an *in vitro* pharmacokinetic/pharmacodynamic model. Antimicrob Agents Chemother 2013; 57(8): 3738-45.
[http://dx.doi.org/10.1128/AAC.00703-13] [PMID: 23716052]

[93] Mauri C, Maraolo AE, Di Bella S, Luzzaro F, Principe L. The Revival of Aztreonam in Combination with Avibactam against Metallo-β-Lactamase-Producing Gram-Negatives: A Systematic Review of *In Vitro* Studies and Clinical Cases. Antibiotics (Basel) 2021; 10(8): 1012.
[http://dx.doi.org/10.3390/antibiotics10081012] [PMID: 34439062]

[94] Shields RK, Doi Y. Aztreonam Combination Therapy: An Answer to Metallo-β-Lactamase–Producing Gram-Negative Bacteria? Clin Infect Dis 2020; 71(4): 1099-101.
[http://dx.doi.org/10.1093/cid/ciz1159] [PMID: 31802110]

[95] Hobson CA, Pierrat G, Tenaillon O, *et al.* *Klebsiella pneumoniae* Carbapenemase Variants Resistant to Ceftazidime-Avibactam: an Evolutionary Overview. Antimicrob Agents Chemother 2022; 66(9): e00447-22.
[http://dx.doi.org/10.1128/aac.00447-22] [PMID: 35980232]

[96] Sagan O, Yakubsevitch R, Yanev K, *et al.* Pharmacokinetics and tolerability of intravenous sulbactam durlobactam with imipenem-cilastatin in hospitalized adults with complicated urinary tract infections, including acute pyelonephritis. Antimicrob Agents Chemother 2020; 64(3): e01506-19.
[http://dx.doi.org/10.1128/AAC.01506-19] [PMID: 31843995]

[97] McLeod SM, O'Donnell JP, Narayanan N, Mills JP, Kaye KS. Sulbactam–durlobactam: a β-lactam/--lactamase inhibitor combination targeting *Acinetobacter baumannii*. Future Microbiol 2024; 19(7): 563-76.
[http://dx.doi.org/10.2217/fmb-2023-0248]

[98] Motaouakkil S, Charra B, Hachimi A, *et al.* Colistin and rifampicin in the treatment of nosocomial infections from multiresistant *Acinetobacter baumannii*. J Infec 2006; 53(4): 274-8.
[http://dx.doi.org/10.1016/j.jinf.2005.11.019]

[99] Bremmer DN, Bauer KA, Pouch SM, *et al.* Correlation of checkerboard synergy testing with time kill analysis and clinical outcomes of extensively drug-resistant *Acinetobacter baumannii* respiratory infections. Antimicrob Agents Chemother 2016; 60(11): 6892-5.
[http://dx.doi.org/10.1128/AAC.00981-16] [PMID: 27527089]

[100] Park HJ, Cho JH, Kim HJ, Han SH, Jeong SH, Byun MK. Colistin monotherapy versus colistin/rifampicin combiation therapy in pneumonia caused by colistin-resistant *Acinetobacter baumannii*: A randomised controlled trial. J Glob Antimicrob Resist 2019; 17: 66-71.
[http://dx.doi.org/10.1016/j.jgar.2018.11.016] [PMID: 30476654]

[101] Song JY, Kee SY, Hwang IS, *et al.* *In vitro* activities of carbapenem/sulbactam combination, colistin, colistin/rifampicin combination and tigecycline against carbapenem-resistant *Acinetobacter baumannii*. J Antimicrob Chemother 2007; 60(2): 317-22.
[http://dx.doi.org/10.1093/jac/dkm136] [PMID: 17540672]

[102] Lim TP, Tan TY, Lee W, *et al.* *In-vitro* activity of polymyxin B, rifampicin, tigecycline alone and in combination against carbapenem-resistant *Acinetobacter baumannii* in Singapore. PLoS One 2011; 6(4): e18485.
[http://dx.doi.org/10.1371/journal.pone.0018485] [PMID: 21533030]

[103] Saballs M, Pujol M, Tubau F, *et al.* Rifampicin/imipenem combination in the treatment of carbapenem-resistant *Acinetobacter baumannii* infections. J Antimicrob Chemother 2006; 58(3): 697-

700.
[http://dx.doi.org/10.1093/jac/dkl274] [PMID: 16895941]

[104] Tamma PD, Cosgrove SE, Maragakis LL. Combination therapy for treatment of infections with gram-negative bacteria. Clin Microbiol Rev 2012; 25(3): 450-70.
[http://dx.doi.org/10.1128/CMR.05041-11] [PMID: 22763634]

[105] O'Donnell JN, Lodise TP. New Perspectives on Antimicrobial Agents: Imipenem-Relebactam. Antimicrob Agents Chemother 2022; 66(7): e00256-22.
[http://dx.doi.org/10.1128/aac.00256-22] [PMID: 35727059]

[106] Lizza BD, Betthauser KD, Ritchie DJ, Micek ST, Kollef MH. New Perspectives on Antimicrobial Agents: Ceftolozane-Tazobactam. Antimicrob Agents Chemother 2021; 65(7): e02318-20.
[http://dx.doi.org/10.1128/AAC.02318-20] [PMID: 33875428]

[107] Baek MS, Chung ES, Jung DS, Ko KS. Effect of colistin-based antibiotic combinations on the eradication of persister cells in *Pseudomonas aeruginosa*. J Antimicrob Chemother 2020; 75(4): 917-24.
[http://dx.doi.org/10.1093/jac/dkz552] [PMID: 31977044]

[108] Crémieux AC, Dinh A, Nordmann P, *et al.* Efficacy of colistin alone and in various combinations for the treatment of experimental osteomyelitis due to carbapenemase-producing *Klebsiella pneumoniae.* J Antimicrob Chemother 2019; 74(9): 2666-75.
[http://dx.doi.org/10.1093/jac/dkz257] [PMID: 31263884]

[109] Alós JI. Resistencia bacteriana a los antibióticos: una crisis global. Enferm Infecc Microbiol Clin 2015; 33(10): 692-9.
[http://dx.doi.org/10.1016/j.eimc.2014.10.004] [PMID: 25475657]

[110] Pang Z, Raudonis R, Glick BR, Lin TJ, Cheng Z. Antibiotic resistance in *Pseudomonas aeruginosa*: mechanisms and alternative therapeutic strategies. Biotechnol Adv 2019; 37(1): 177-92.
[http://dx.doi.org/10.1016/j.biotechadv.2018.11.013] [PMID: 30500353]

[111] East SP, Silver LL. Multitarget ligands in antibacterial research: progress and opportunities. Expert Opin Drug Discov 2013; 8(2): 143-56.
[http://dx.doi.org/10.1517/17460441.2013.743991] [PMID: 23252414]

[112] Gray DA, Wenzel M. Multitarget Approaches against Multiresistant Superbugs. ACS Infect Dis 2020; 6(6): 1346-65.
[http://dx.doi.org/10.1021/acsinfecdis.0c00001] [PMID: 32156116]

[113] Liu Y, Li R, Xiao X, Wang Z. Antibiotic adjuvants: an alternative approach to overcome multi-drug resistant Gram-negative bacteria. Crit Rev Microbiol 2019; 45(3): 301-14.
[http://dx.doi.org/10.1080/1040841X.2019.1599813] [PMID: 30985240]

[114] Tyers M, Wright GD. Drug combinations: a strategy to extend the life of antibiotics in the 21st century. Nat Rev Microbiol 2019; 17(3): 141-55.
[http://dx.doi.org/10.1038/s41579-018-0141-x] [PMID: 30683887]

[115] Gupta V, Datta P. Next-generation strategy for treating drug resistant bacteria. Indian J Med Res 2019; 149(2): 97-106.
[http://dx.doi.org/10.4103/ijmr.IJMR_755_18] [PMID: 31219074]

[116] Cottarel G, Wierzbowski J. Combination drugs, an emerging option for antibacterial therapy. Trends Biotechnol 2007; 25(12): 547-55.
[http://dx.doi.org/10.1016/j.tibtech.2007.09.004] [PMID: 17997179]

[117] Scaiola A, Leibundgut M, Boehringer D, *et al.* Structural basis of translation inhibition by cadazolid, a novel quinoxolidinone antibiotic. Sci Rep 2019; 9(1): 5634.
[http://dx.doi.org/10.1038/s41598-019-42155-4] [PMID: 30948752]

[118] "Idorsia announces financial results for the first quarter 2018". Idorsia. April 19, 2018. Archived from the original on April 28, 2018. Retrieved April 27,2018. https://www.idorsia.com/media/news-

details?newsId=1652505.

[119] Dhiman S, Ramirez D, Li Y, Kumar A, Arthur G, Schweizer F. Chimeric tobramycin-based adjuvant TOB-TOB-CIP potentiates fluoroquinolone and β-lactam antibiotics against multidrug-resistant *Pseudomonas aeruginosa*. ACS Infect Dis 2023; 9(4): 864-85.
[http://dx.doi.org/10.1021/acsinfecdis.2c00549] [PMID: 36917096]

[120] Stryjewski ME, Potgieter PD, Li YP, *et al.* TD-1792 versus vancomycin for treatment of complicated skin and skin structure infections. Antimicrob Agents Chemother 2012; 56(11): 5476-83.
[http://dx.doi.org/10.1128/AAC.00712-12] [PMID: 22869571]

[121] Ma Z, Lynch AS. Development of a Dual-Acting Antibacterial Agent (TNP-2092) for the Treatment of Persistent Bacterial Infections J Med Chem 2016; 59(14): 6645-57.

[122] Xiao ZP, Wang XD, Wang PF, *et al.* Design, synthesis, and evaluation of novel fluoroquinolone–flavonoid hybrids as potent antibiotics against drug-resistant microorganisms. Eur J Med Chem 2014; 80: 92-100.
[http://dx.doi.org/10.1016/j.ejmech.2014.04.037] [PMID: 24769347]

[123] Tonziello G, Caraffa E, Pinchera B, Granata G, Petrosillo N. Present and future of siderophore-based therapeutic and diagnostic approaches in infectious diseases. Infect Dis Rep 2019; 11(2): 8208.
[http://dx.doi.org/10.4081/idr.2019.8208] [PMID: 31649808]

[124] El-Lababidi RM, Rizk JG. Cefiderocol: A Siderophore Cephalosporin. Ann Pharmacother 2020; 54(12): 1215-31.
[http://dx.doi.org/10.1177/1060028020929988] [PMID: 32522005]

[125] Gehrmann R, Hertlein T, Hopke E, Ohlsen K, Lalk M, Hilgeroth A. Novel small-molecule hybrid-antibacterial agents against *S. aureus* and MRSA strains. Molecules 2021; 27(1): 61.
[http://dx.doi.org/10.3390/molecules27010061] [PMID: 35011293]

[126] Schaefer AL, Hanzelka BL, Eberhard A, Greenberg EP. Quorum sensing in *Vibrio fischeri*: probing autoinducer-LuxR interactions with autoinducer analogs. J Bacteriol 1996; 178(10): 2897-901.
[http://dx.doi.org/10.1128/jb.178.10.2897-2901.1996] [PMID: 8631679]

[127] Duplantier M, Lohou E, Sonnet P. Quorum Sensing Inhibitors to Quench *P. aeruginosa* Pathogenicity. Pharmaceuticals (Basel) 2021; 14(12): 1262.
[http://dx.doi.org/10.3390/ph14121262] [PMID: 34959667]

[128] Schütz C, Empting M. Targeting the *Pseudomonas* quinolone signal quorum sensing system for the discovery of novel anti-infective pathoblockers. Beilstein J Org Chem 2018; 14: 2627-45.
[http://dx.doi.org/10.3762/bjoc.14.241] [PMID: 30410625]

[129] Rampioni G, Falcone M, Heeb S, *et al.* Unravelling the Genome-Wide Contributions of Specific 2-Alkyl-4-Quinolones and PqsE to Quorum Sensing in *Pseudomonas aeruginosa*. PLoS Pathog 2016; 12(11): e1006029.
[http://dx.doi.org/10.1371/journal.ppat.1006029] [PMID: 27851827]

[130] Welsh MA, Blackwell HE. Chemical Genetics Reveals Environment-Specific Roles for Quorum Sensing Circuits in *Pseudomonas aeruginosa*. Cell Chem Biol 2016; 23(3): 361-9.
[http://dx.doi.org/10.1016/j.chembiol.2016.01.006] [PMID: 26905657]

[131] Maura D, Drees SL, Bandyopadhaya A, *et al.* Polypharmacology Approaches against the *Pseudomonas aeruginosa* MvfR Regulon and Their Application in Blocking Virulence and Antibiotic Tolerance. ACS Chem Biol 2017; 12(5): 1435-43.
[http://dx.doi.org/10.1021/acschembio.6b01139] [PMID: 28379691]

[132] Lee J, Zhang L. The hierarchy quorum sensing network in *Pseudomonas aeruginosa*. Protein Cell 2015; 6(1): 26-41.
[http://dx.doi.org/10.1007/s13238-014-0100-x] [PMID: 25249263]

[133] Kostylev M, Kim DY, Smalley NE, Salukhe I, Greenberg EP, Dandekar AA. Evolution of the *Pseudomonas aeruginosa* quorum-sensing hierarchy. Proc Natl Acad Sci USA 2019; 116(14): 7027-

32.
[http://dx.doi.org/10.1073/pnas.1819796116] [PMID: 30850547]

[134] Malgaonkar A, Nair M. Quorum sensing in Pseudomonas aeruginosa mediated by RhlR is regulated by a small RNA PhrD Sci Rep 2019; 9: 432.
[http://dx.doi.org/10.1038/s41598-018-36488-9]

[135] Le KY, Otto M. Quorum-sensing regulation in *Staphylococci*—an overview. Front Microbiol 2015; 6: 1174.
[http://dx.doi.org/10.3389/fmicb.2015.01174] [PMID: 26579084]

[136] Surette MG, Miller MB, Bassler BL. Quorum sensing in *Escherichia coli, Salmonella typhimurium*, and *Vibrio harveyi* : A new family of genes responsible for autoinducer production. Proc Natl Acad Sci USA 1999; 96(4): 1639-44.
[http://dx.doi.org/10.1073/pnas.96.4.1639] [PMID: 9990077]

[137] Coquant G, Grill JP, Seksik P. Impact of *N*-Acyl-Homoserine Lactones, Quorum Sensing Molecules, on Gut Immunity. Front Immunol 2020; 11: 1827.
[http://dx.doi.org/10.3389/fimmu.2020.01827] [PMID: 32983093]

[138] Styles MJ, Early SA, Tucholski T, West KHJ, Ge Y, Blackwell HE. Chemical Control of Quorum Sensing in *E. coli*: Identification of Small Molecule Modulators of SdiA and Mechanistic Characterization of a Covalent Inhibitor. ACS Infect Dis 2020; 6(12): 3092-103.
[http://dx.doi.org/10.1021/acsinfecdis.0c00654] [PMID: 33124430]

[139] Brameyer S, Bode HB, Heermann R. Languages and dialects: bacterial communication beyond homoserine lactones. Trends Microbiol 2015; 23(9): 521-3.
[http://dx.doi.org/10.1016/j.tim.2015.07.002] [PMID: 26231578]

[140] Trottier MC, de Oliveira Pereira T, Groleau MC, Hoffman LR, Dandekar AA, Déziel E. The end of the reign of a "master regulator"? A defect in function of the LasR quorum sensing regulator is a common feature of *Pseudomonas aeruginosa* isolates. MBio 2024; 15(3): e02376-23.
[http://dx.doi.org/10.1128/mbio.02376-23] [PMID: 38315035]

[141] Martin CA, Hoven AD, Cook AM. Therapeutic frontiers: preventing and treating infectious diseases by inhibiting bacterial quorum sensing. Eur J Clin Microbiol Infect Dis 2008; 27(8): 635-42.
[http://dx.doi.org/10.1007/s10096-008-0489-3] [PMID: 18322716]

[142] Nalca Y, Jänsch L, Bredenbruch F, Geffers R, Buer J, Häussler S. Quorum-sensing antagonistic activities of azithromycin in *Pseudomonas aeruginosa* PAO1: a global approach. Antimicrob Agents Chemother 2006; 50(5): 1680-8.
[http://dx.doi.org/10.1128/AAC.50.5.1680-1688.2006] [PMID: 16641435]

[143] Kai T, Tateda K, Kimura S, *et al.* A low concentration of azithromycin inhibits the mRNA expression of N-acyl homoserine lactone synthesis enzymes, upstream of lasI or rhlI, in *Pseudomonas aeruginosa*. Pulm Pharmacol Ther 2009; 22(6): 483-6.
[http://dx.doi.org/10.1016/j.pupt.2009.04.004] [PMID: 19393329]

[144] Imamura Y, Higashiyama Y, Tomono K, *et al.* Azithromycin exhibits bactericidal effects on *Pseudomonas aeruginosa* through interaction with the outer membrane. Antimicrob Agents Chemother 2005; 49(4): 1377-80.
[http://dx.doi.org/10.1128/AAC.49.4.1377-1380.2005] [PMID: 15793115]

[145] Leroy AG, Caillon J, Caroff N, *et al.* Could azithromycin be part of *Pseudomonas aeruginosa* acute pneumonia treatment? Front Microbiol 2021; 12: 642541.
[http://dx.doi.org/10.3389/fmicb.2021.642541] [PMID: 33796090]

[146] Murray EJ, Dubern JF, Chan WC, Chhabra SR, Williams P. A *Pseudomonas aeruginosa* PQS quorum-sensing system inhibitor with anti-staphylococcal activity sensitizes polymicrobial biofilms to tobramycin. Cell Chem Biol 2022; 29(7): 1187-1199.e6.
[http://dx.doi.org/10.1016/j.chembiol.2022.02.007] [PMID: 35259345]

[147] Soo V, Kwan B, Quezada H, *et al.* Repurposing of Anticancer Drugs for the Treatment of Bacterial Infections. Curr Top Med Chem 2017; 17(10): 1157-76.
[http://dx.doi.org/10.2174/1568026616666160930131737] [PMID: 27697046]

[148] D'Angelo F, Baldelli V, Halliday N, *et al.* Identification of FDA-Approved Drugs as Antivirulence Agents Targeting the *pqs* Quorum-Sensing System of *Pseudomonas aeruginosa.* Antimicrob Agents Chemother 2018; 62(11): e01296-18.
[http://dx.doi.org/10.1128/AAC.01296-18] [PMID: 30201815]

[149] Khan F, Pham DTN, Oloketuyi SF, Kim YM. Regulation and controlling the motility properties of *Pseudomonas aeruginosa.* Appl Microbiol Biotechnol 2020; 104(1): 33-49.
[http://dx.doi.org/10.1007/s00253-019-10201-w] [PMID: 31768614]

[150] Dusane DH, Zinjarde SS, Venugopalan VP, Mclean RJC, Weber MM, Rahman PKSM. Quorum sensing: implications on Rhamnolipid biosurfactant production. Biotechnol Genet Eng Rev 2010; 27(1): 159-84.
[http://dx.doi.org/10.1080/02648725.2010.10648149] [PMID: 21415897]

[151] Chang CY, Krishnan T, Wang H, *et al.* Non-antibiotic quorum sensing inhibitors acting against N-acyl homoserine lactone synthase as druggable target. Sci Rep 2014; 4(1): 7245.
[http://dx.doi.org/10.1038/srep07245] [PMID: 25430794]

[152] Topa SH, Palombo EA, Kingshott P, Blackall LL. Activity of Cinnamaldehyde on Quorum Sensing and Biofilm Susceptibility to Antibiotics in *Pseudomonas aeruginosa.* Microorganisms 2020; 8(3): 455.
[http://dx.doi.org/10.3390/microorganisms8030455] [PMID: 32210139]

[153] Tapia-Rodriguez MR, Bernal-Mercado AT, Gutierrez-Pacheco MM, *et al.* Virulence of *Pseudomonas aeruginosa* exposed to carvacrol: alterations of the Quorum sensing at enzymatic and gene levels. J Cell Commun Signal 2019; 13(4): 531-7.
[http://dx.doi.org/10.1007/s12079-019-00516-8] [PMID: 30903602]

[154] Grandclément C, Tannières M, Moréra S, Dessaux Y, Faure D. Quorum quenching: role in nature and applied developments. FEMS Microbiol Rev 2016; 40(1): 86-116.
[http://dx.doi.org/10.1093/femsre/fuv038] [PMID: 26432822]

[155] Kalia VC, Raju C, Purohit HJ. Genomic Analysis Reveals Versatile Organisms for Quorum Quenching Enzymes: Acyl-Homoserine Lactone-Acylase and -Lactonase. The Open Microbiology Journal 2011; 5: 1-13.

[156] Dong YH, Wang LH, Zhang LH. Quorum-quenching microbial infections: mechanisms and implications. Philos Trans R Soc Lond B Biol Sci 2007; 362(1483): 1201-11.
[http://dx.doi.org/10.1098/rstb.2007.2045] [PMID: 17360274]

[157] Persson T, Givskov M, Nielsen J. Quorum sensing inhibition: targeting chemical communication in gram-negative bacteria. Curr Med Chem 2005; 12(26): 3103-15.
[http://dx.doi.org/10.2174/092986705774933425] [PMID: 16375704]

[158] Castang S, Chantegrel B, Deshayes C, *et al.* N-Sulfonyl homoserine lactones as antagonists of bacterial quorum sensing. Bioorg Med Chem Lett 2004; 14(20): 5145-9.
[http://dx.doi.org/10.1016/j.bmcl.2004.07.088] [PMID: 15380216]

[159] Li SZ, Xu R, Ahmar M, Goux-Henry C, Queneau Y, Soulère L. Influence of the d/l configuration of N-acyl-homoserine lactones (AHLs) and analogues on their Lux-R dependent quorum sensing activity. Bioorg Chem 2018; 77: 215-22.
[http://dx.doi.org/10.1016/j.bioorg.2018.01.005] [PMID: 29367078]

[160] Ishida T, Ikeda T, Takiguchi N, Kuroda A, Ohtake H, Kato J. Inhibition of quorum sensing in *Pseudomonas aeruginosa* by N-acyl cyclopentylamides. Appl Environ Microbiol 2007; 73(10): 3183-8.
[http://dx.doi.org/10.1128/AEM.02233-06] [PMID: 17369333]

[161] Morohoshi T, Shiono T, Takidouchi K, *et al.* Inhibition of quorum sensing in Serratia marcescens AS-1 by synthetic analogs of N-acylhomoserine lactone. Appl Environ Microbiol 2007; 73(20): 6339-44.
[http://dx.doi.org/10.1128/AEM.00593-07] [PMID: 17675425]

[162] Frezza M, Soulère L, Reverchon S, *et al.* Synthetic homoserine lactone-derived sulfonylureas as inhibitors of *Vibrio fischeri* quorum sensing regulator. Bioorg Med Chem 2008; 16(7): 3550-6.
[http://dx.doi.org/10.1016/j.bmc.2008.02.023] [PMID: 18294853]

[163] Rasmussen TB, Bjarnsholt T, Skindersoe ME, *et al.* Screening for quorum-sensing inhibitors (QSI) by use of a novel genetic system, the QSI selector. J Bacteriol 2005; 187(5): 1799-814.
[http://dx.doi.org/10.1128/JB.187.5.1799-1814.2005] [PMID: 15716452]

[164] Vanoyan N, Walker SL, Gillor O, Herzberg M. Reduced bacterial deposition and attachment by quorum-sensing inhibitor 4-nitro-pyridine-N-oxide: the role of physicochemical effects. Langmuir 2010; 26(14): 12089-94.
[http://dx.doi.org/10.1021/la101319e] [PMID: 20553026]

[165] Zhang Q, Li S, Hachicha M, *et al.* Heterocyclic Chemistry Applied to the Design of *N*-Acyl Homoserine Lactone Analogues as Bacterial Quorum Sensing Signals Mimics. Molecules 2021; 26(17): 5135.
[http://dx.doi.org/10.3390/molecules26175135] [PMID: 34500565]

[166] Allegretta G, Maurer CK, Eberhard J, *et al.* In-depth Profiling of MvfR-Regulated Small Molecules in *Pseudomonas aeruginosa* after Quorum Sensing Inhibitor Treatment. Front Microbiol 2017; 8: 924.
[http://dx.doi.org/10.3389/fmicb.2017.00924] [PMID: 28596760]

[167] Hentzer M, Riedel K, Rasmussen TB, *et al.* Inhibition of quorum sensing in *Pseudomonas aeruginosa* biofilm bacteria by a halogenated furanone compound. Microbiology (Reading) 2002; 148(1): 87-102.
[http://dx.doi.org/10.1099/00221287-148-1-87] [PMID: 11782502]

[168] Rasmussen TB, Manefield M, Andersen JB, *et al.* How Delisea pulchra furanones affect quorum sensing and swarming motility in *Serratia liquefaciens* MG1. Microbiology (Reading) 2000; 146(12): 3237-44.
[http://dx.doi.org/10.1099/00221287-146-12-3237] [PMID: 11101681]

[169] Manefield M, Rasmussen TB, Henzter M, *et al.* Halogenated furanones inhibit quorum sensing through accelerated LuxR turnover. Microbiology (Reading) 2002; 148(4): 1119-27.
[http://dx.doi.org/10.1099/00221287-148-4-1119] [PMID: 11932456]

[170] Steenackers HP, Levin J, Janssens JC, *et al.* Structure–activity relationship of brominated 3-alkyl-5-methylene-2(5H)-furanones and alkylmaleic anhydrides as inhibitors of *Salmonella biofilm* formation and quorum sensing regulated bioluminescence in Vibrio harveyi. Bioorg Med Chem 2010; 18(14): 5224-33.
[http://dx.doi.org/10.1016/j.bmc.2010.05.055] [PMID: 20580562]

[171] Lönn-Stensrud J, Landin MA, Benneche T, Petersen FC, Scheie AA. Furanones, potential agents for preventing *Staphylococcus epidermidis* biofilm infections? J Antimicrob Chemother 2008; 63(2): 309-16.
[http://dx.doi.org/10.1093/jac/dkn501] [PMID: 19098295]

[172] Bové M, Bao X, Sass A, Crabbé A, Coenye T. The quorum-sensing inhibitor furanone C-30 rapidly loses its tobramycin-potentiating activity against *Pseudomonas aeruginosa* biofilms during experimental evolution. Antimicrob Agents Chemother 2021; 65(7): e00413-21.
[http://dx.doi.org/10.1128/AAC.00413-21] [PMID: 33903100]

[173] Muñoz-Cázares N, Castillo-Juárez I, García-Contreras R, *et al.* A Brominated Furanone Inhibits *Pseudomonas aeruginosa* Quorum Sensing and Type III Secretion, Attenuating Its Virulence in a Murine Cutaneous Abscess Model. Biomedicines 2022; 10(8): 1847.
[http://dx.doi.org/10.3390/biomedicines10081847] [PMID: 36009394]

[174] Gajdács M, Baráth Z, Kárpáti K, *et al.* No Correlation between Biofilm Formation, Virulence Factors,

and Antibiotic Resistance in *Pseudomonas aeruginosa*: Results from a Laboratory-Based In Vitro Study. Antibiotics (Basel) 2021; 10(9): 1134.
[http://dx.doi.org/10.3390/antibiotics10091134] [PMID: 34572716]

[175] Hibbing ME, Fuqua C, Parsek MR, Peterson SB. Bacterial competition: surviving and thriving in the microbial jungle. Nat Rev Microbiol 2010; 8(1): 15-25.
[http://dx.doi.org/10.1038/nrmicro2259] [PMID: 19946288]

[176] Uroz S, Dessaux Y, Oger P. Quorum sensing and quorum quenching: the yin and yang of bacterial communication. ChemBioChem 2009; 10(2): 205-16.
[http://dx.doi.org/10.1002/cbic.200800521] [PMID: 19072824]

[177] El-Ghali A, Kunz Coyne AJ, Caniff K, Bleick C, Rybak MJ. Sulbactam-durlobactam: A novel β-lactam-β-lactamase inhibitor combination targeting carbapenem resistant *Acinetobacter baumannii* infections. Pharmacotherapy 2023; 43(6): 502-13.
[http://dx.doi.org/10.1002/phar.2802] [PMID: 37052117]

[178] Domalaon R, Idowu T, Zhanel GG, Schweizer F. Antibiotic Hybrids: the Next Generation of Agents and Adjuvants against Gram-Negative Pathogens? Clin Microbiol Rev 2018; 31(2): e00077-17.
[http://dx.doi.org/10.1128/CMR.00077-17] [PMID: 29540434]

[179] Andrei S, Droc G, Stefan G. FDA approved antibacterial drugs: 2018-2019. Discoveries (Craiova) 2019; 7(4): e102. https://discoveriesjournals.org/discoveries/D.2019.04.FR-Andrei.DOI
[http://dx.doi.org/10.15190/d.2019.15] [PMID: 32309620]

[180] Dhiman S, Ramirez D, Li Y, Kumar A, Arthur G, Schweizer F. Chimeric Tobramycin-Based Adjuvant TOB-TOB-CIP Potentiates Fluoroquinolone and β-Lactam Antibiotics against Multidrug-Resistant *Pseudomonas aeruginosa*. ACS Infect Dis 2023; 9(4): 864-85.
[http://dx.doi.org/10.1021/acsinfecdis.2c00549] [PMID: 36917096]

[181] Dhiman S, Ramirez D, Arora R, *et al.* Trimeric Tobramycin/Nebramine Synergizes β-Lactam Antibiotics against *Pseudomonas aeruginosa*. ACS Omega 2023; 8(32): 29359-73.
[http://dx.doi.org/10.1021/acsomega.3c02810] [PMID: 37599980]

[182] Rudramurthy G, Swamy M, Sinniah U, Ghasemzadeh A. Nanoparticles: Alternatives Against Drug-Resistant Pathogenic Microbes. Molecules 2016; 21(7): 836.
[http://dx.doi.org/10.3390/molecules21070836] [PMID: 27355939]

[183] Akhtar MS, Swamy MK, Umar A, Al Sahli AA. Biosynthesis and characterization of silver nanoparticles from methanol leaf extract of Cassia didymobotyra and assessment of their antioxidant and antibacterial activities. J Nanosci Nanotechnol 2015; 15(12): 9818-23.
[http://dx.doi.org/10.1166/jnn.2015.10966] [PMID: 26682418]

[184] Zhao X, Jia Y, Dong R, *et al.* Bimetallic nanoparticles against multi-drug resistant bacteria. Chem Commun (Camb) 2020; 56(74): 10918-21.
[http://dx.doi.org/10.1039/D0CC03481A] [PMID: 32808607]

[185] Synergistic effect of antibiotic with green synthesized silver nanoparticles against uropathogenic E, coli biofilm. Muna H., Alshaikhly NS., Al-khafaj M. H.M. Iraqi Journal of Agricultural Sciences 2023; 54(6): 1622-35.
[http://dx.doi.org/10.36103/ijas.v54i6.1862]

[186] Mahapatro, A., Singh, D.K. Biodegradable nanoparticles are excellent vehicle for site directed in-vivo delivery of drugs and vaccines. J Nanobiotechnol 2011; 9: 55.

[187] Kumari A, Yadav SK, Yadav SC. Biodegradable polymeric nanoparticles based drug delivery systems. Colloids Surf B Biointerfaces 2010; 75(1): 1-18.
[http://dx.doi.org/10.1016/j.colsurfb.2009.09.001] [PMID: 19782542]

[188] Abo-zeid Y, Amer A, Bakkar MR, El-Houssieny B, Sakran W. Antimicrobial Activity of Azithromycin Encapsulated into PLGA NPs: A Potential Strategy to Overcome Efflux Resistance. Antibiotics (Basel) 2022; 11(11): 1623.

[http://dx.doi.org/10.3390/antibiotics11111623] [PMID: 36421266]

[189] Vibe CB, Fenaroli F, Pires D, *et al.* Thioridazine in PLGA nanoparticles reduces toxicity and improves rifampicin therapy against mycobacterial infection in zebrafish. Nanotoxicology 2016; 10(6): 680-8.
[http://dx.doi.org/10.3109/17435390.2015.1107146] [PMID: 26573343]

[190] Lewies A, Wentzel JF, Jordaan A, Bezuidenhout C, Du Plessis LH. Interactions of the antimicrobial peptide nisin Z with conventional antibiotics and the use of nanostructured lipid carriers to enhance antimicrobial activity. Int J Pharm 2017; 526(1-2): 244-53.
[http://dx.doi.org/10.1016/j.ijpharm.2017.04.071] [PMID: 28461263]

[191] Xie S, Tao Y, Pan Y, *et al.* Biodegradable nanoparticles for intracellular delivery of antimicrobial agents. J Control Release 2014; 187: 101-17.
[http://dx.doi.org/10.1016/j.jconrel.2014.05.034] [PMID: 24878179]

[192] Lacoma A, Usón L, Mendoza G, *et al.* Novel intracellular antibiotic delivery system against *Staphylococcus aureus*: cloxacillin-loaded poly(d,l-lactide-co-glycolide) acid nanoparticles. Nanomedicine (Lond) 2020; 15(12): 1189-203.
[http://dx.doi.org/10.2217/nnm-2019-0371] [PMID: 32370602]

SUBJECT INDEX

www.ingramcontent.com/pod-product-compliance
Lightning Source LLC
Chambersburg PA
CBHW050838220326
41598CB00006B/394